Applied Digital Logic Exercises Using FPGAs

Applied Digital Logic Exercises Using FPGAs

Kurt Wick
University of Minnesota, USA

Morgan & Claypool Publishers

ISBN   978-1-6817-4660-9 (ebook)
ISBN   978-1-6817-4661-6 (print)
ISBN   978-1-6817-4662-3 (mobi)

DOI   10.1088/978-1-6817-4660-9

Version: 20171001

IOP Concise Physics
ISSN 2053-2571 (online)
ISSN 2054-7307 (print)

A Morgan & Claypool publication as part of IOP Concise Physics
Published by Morgan & Claypool Publishers, 1210 Fifth Avenue, Suite 250, San Rafael, CA, 94901, USA

IOP Publishing, Temple Circus, Temple Way, Bristol BS1 6HG, UK

To Ying, Chloe and Abby.

Contents

Preface

This book is for anyone interested in digital logic and who wants to learn how to implement it through detailed exercises with state of the art digital design tools and components. It exposes the reader to combinational and sequential digital logic concepts and implements them with hands-on exercises using the Verilog hardware description language (HDL) and a field programmable gate arrays (FGPA) teaching board.

This book covers basic digital design concepts and then applies them through exercises. It teaches the reader the syntax of the Verilog language to build a simple calculator, a basic music player, a frequency and period counter and it ends with a microprocessor being embedded in the fabric of the FGPA to communicate with the PC. In the process, the reader learns about digital mathematics and digital-to-analog converter concepts through pulse width modulation.

The topics are aimed at an audience of undergraduate students in a science or engineering course (or hobbyists) wanting an introduction to applied digital logic concepts. The material is deliberately kept brief so it can be covered in about four (course) weeks. No prior knowledge of digital logic or the Verilog programming language is required.

Hardware & Software Requirements:
Verilog compiler software, Vivado Version 2016.x., can be obtained (for free) by Xilinx Inc., and FPGA teaching boards, BASYS3 by Digilent Inc. are sold for about $100 making them affordable for students and hobbyists. Access to an oscilloscope and a function generator is required for some of the exercises. Detailed software installation instructions are presented in appendix A.

Acknowledgments

I extend my thanks to Kevin Booth and Professor Müller for their constructive input and proofreading. I would also like to thank Adam Schaefer, Luke Molacek and Siddarth Karuka for finding many, but by no means all, tedious mistakes in this manual. Thanks also to James Duckworth at Worcester Polytechnic Institute for his documents and help in interfacing the FPGA with the microprocessor, and to Jeremiah Mans for providing the code for the binary to BCD conversion for the HEX display. Last but not least, thanks to all the Fall 2016 Phys4051 students who worked through the exercises and provided feedback.

Kurt Wick, University of Minnesota, 13 June 2017

Author biography

Kurt Wick

Kurt Wick received his MS in 1989 from the University of Minnesota where he has been developing and teaching, as Senior Scientist, the methods of experimental physics advanced laboratory course. The course covers a wide range of fields, such as optics, solid state and high energy physics and exposes students to electronics, programing and computer interfacing, statistics and data analysis.

With his background in computers and electronics he is interested in teaching the fundamentals of digital electronics and how it pertains to today's consumer electronics and lab instruments. For the last five years, he has presented workshops on FPGAs through the Advanced Laboratory in Physics Association and when he's not working, he enjoys paddling in the woods of northern Minnesota and Ontario.

Applied Digital Logic Exercises Using FPGAs

Kurt Wick

Chapter 1

Introduction to digital logic

Additional reading

Read pages 703–24 in Horowitz P and Hill W 2015 *The Art of Electronics* 3rd edn (New York: Cambridge University Press) and pages 717–37 in Scherz P and Monk S 2013 *Practical Electronics for Inventors* 3rd edn (New York: McGraw-Hill Education).

1.1 Basic definitions of digital concepts

1.1.1 Definition

While an analog signal is continuous, a digital one is discrete. Specifically, an analog signal represents an infinite continuum of levels while the digital one represents finite, discrete levels.

1.1.2 Analog versus digital analogy

Our world is inherently analog. We sense continuous signal levels such as the intensity of light or sound. In contrast, if humans were born with digital sensory organs we would only sense binary states, for example brightness or darkness, silence or noise. Not surprisingly, most of us who are fortunate to enjoy functioning analog sensory organs would never want to trade these for digital ones. Based on this observation it appears that in comparison to analog, digital technology is inherently limited and crude. Nevertheless, the prevailing trend in technology clearly favors the digital approach over the 'old' analog one, indicating significant advantages of this 'new' technology. Let's explore the main advantages with the following analogy.

Imagine that you are in a concert hall with two 'musicians' who are hidden from the audience by a curtain. The task for the performers is simple: they are each asked to play a single note on their instrument and then the audience is asked to describe the sound intensity or if they even heard that instrument was played.

After the first performer plays a few identical notes on a violin you probably would get a wide response from the listeners ranging from 'quiet' to 'loud'. After all, the perception is subjective and based on the listener's music preference, hearing capability and distance from the player. Furthermore, depending on the relationship between how loud the note was played and how loud the background noise was in the hall, it may even be hard for the listeners to discern the state of the instrument, i.e. was it played at all or did they only hear noise.

Next imagine the second player, who uses an extremely loud and obnoxious sounding car horn as his instrument. Again he plays one note by turning the horn on or off multiple times. Now, most listeners would clearly agree on the state of the perceived sound intensity, i.e. either it was quiet or loud—the horn was 'on' or 'off'. As long as the car horn is louder than the background noise level in the concert hall (a very likely condition) no ambiguity of the signal state exits.

If our only concern is to make certain that the audience perceives the signal's state, i.e. is the instrument being played or not, then clearly the second musician with his digital approach wins. By using an instrument capable of only producing two clearly discernable states, complete silence and ear piercing noise, he is able to transmit his message across the concert hall without ambiguity or errors. It should come as no surprise why this digital message system works so well; after all, it was designed to work in an extremely noisy environment, such as traffic, to deliver a clear (warning) message.

In summary, by only using clearly identifiable states, digital technology is able to transmit a signal state that is (almost) immune to noise. By the same reasoning, it follows that these states can also be duplicated and stored without degradation by ambient noise. In digital technology, such a binary 'signal' is referred to as a 'bit' and its state as 'on' or 'off'.

Of course, one could make the argument that we are really comparing two completely different systems in our concert hall analogy. It is one thing to convey a clear message by, for example, either remaining completely quiet or blasting an obnoxious car horn. However, it is a completely different matter to convey a more complex, nuanced and probably esthetically pleasing message, such as the violinist may produce with her analog instrument. Instead, the question we should ask is: can we produce a more nuanced message with our simple car horn? At first, this may seem impossible if we are only allowed to use the two clearly discernable states that our car horn emits, silence and noise. However, the digital world gets around this issue by 'bundling' bits into a package and then assigning meaning to this sequence of bits. This is actually not all that different from the way some drivers operate their car horn: no honking means you are ok; one brief honk means you have mildly irritated someone; a long series of honks clearly means you should get off your phone. While this is a bit oversimplified, it shows that we can express a discrete level of emotions by using only silence and no silence.

It also highlights another aspect of digital technology, namely that the more signals, or bits, we agree to use in our sequence, the finer we can express these levels. However, as long as we are forced to use a finite sequence of bits, the number of levels we can express will also be finite. Going back to our traffic analogy, while with one bit we can only express two levels, 'ok' or 'angry', with just two bits we can signal four levels, 'ok', 'mildly irritated', 'somewhat irritated' and 'angry'.

Another advantage that digital technology has over analog is the ease with which the data can be manipulated using very simple 'operators', called *gates*, to mimic mathematical operations.

In summary, the high noise immunity of transmitting and storing digital signals and the ease with which they can be manipulated with mathematical operations give digital an advantage over analog. This explains the current trend away from analog and towards digital technology. Nevertheless, you should keep in mind that we started out by saying that our world (including us humans) is inherently analog. Therefore, at least as long as humans have no digital ports, we will always be working with both analog and digital technology and the interface between these two worlds, analog-to-digital and digital-to-analog converters.

1.1.3 Binary states and digital logic levels

When we refer to the binary state of a bit we will use the conventional labels of 'on' or 'off', '1' or '0' or 'high' or 'low' (which is sometimes also written as 'HI' or 'LO'). We will use these labels interchangeably and agree that 'on', '1' and 'high' are considered equivalent and so are 'off', '0' and 'low'.

In a physical environment, such as an electronic circuit, explicit voltage or current level standards exist for a specific digital logic state. Multiple conventions exist and the most common ones are listed in table 1.1. To conserve energy and to increase switching speeds, the 3.3 V low voltage transistor–transistor logic (TTL) standard is increasingly replacing the old, established 5 V TTL level.

Table 1.1 is deliberately kept simple and it does not list the acceptable threshold levels for high and low input and output signals. Typically these are within one volt of the 'ideal' levels listed above. For example, for the 3.3 V CMOS systems, an input voltage between 0 and 0.8 V is acceptable as a low input, signals between 2.0 and 3.3 V are considered as a high input; you want to avoid input signal levels between 0.8 V and 2.0 V because they are ambiguous.

Discrete components require external power and the levels listed above state their supply voltages.

Finally, avoid applying input signals exceeding the high or low levels stated above. Doing so can destroy the components!

1.1.4 Manipulating digital logic levels with operators or gates

As mentioned previously, the power of digital logic comes from the ability to manipulate the signals through logic gates. These logic gates are to digital technology what operators are to mathematics. There are only three fundamental gates: the NOT, AND and OR gates. All other gates and the entirety of digital technology are based on repeatedly applying these three operators in various combinations. While this may be hard to believe, remember that a large part of mathematics is composed of repeated applications of the addition and subtraction

Table 1.1. Most commonly used digital voltage standards.

Logic level	3.3 V CMOS OR TTL	5 V TTL
1, on, high	3.3 V	5 V
0, off, low	0 V	0 V

operator, resulting essentially in multiplication and division. As you will see later, by combining these gates you will be able to build a simple calculator, a music player and even a microprocessor.

1.2 Digital and Boolean logic and its representation

Multiple methods exist to represent digital logic, each having its strengths and weaknesses. Working with digital designs requires the ability to move fluently from one representation to the next.

All the basic gates and four corresponding representations are shown in table 1.2. In these examples, A and B are inputs and Q is the output.

Table 1.2. Table of equivalent digital logic representations.

Name(s)	Schematic representation	Truth table			Boolean notation	Verilog notation
	Group I					
NOT (inverter)		A 0 1	Q 1 0		$Q = \bar{A}$ $Q = {\sim}A$ $Q = !A$ (sloppy)	assign Q = ~A;
OR (ANY)		A 0 0 1 1	B 0 1 0 1	Q 0 1 1 1	$Q = A + B$	assign Q = A|B;
AND (ALL)		A 0 0 1 1	B 0 1 0 1	Q 0 0 0 1	$Q = AB$ $Q = A{\cdot}B$	assign Q = A&B;
	Group II					
NOR		A 0 0 1 1	B 0 1 0 1	Q 1 0 0 0	$Q = \overline{A + B}$	assign Q = ~(A|B);
NAND		A 0 0 1 1	B 0 1 0 1	Q 1 1 1 0	$Q = \overline{AB}$	assign Q = ~(A&B);
	Group III					
XOR 'exclusive OR'		A 0 0 1 1	B 0 1 0 1	Q 0 1 1 0	$Q = A \oplus B$	assign Q = A^B;

We augmented the three fundamental gates in group I with three additional gates. The gates in group II, the NOR and NAND gates, are a combination of either an OR or an AND gate with a NOT gate. The XOR gate in group III is a combination of all three fundamental gates. These additional three gates are used so frequently that they have been awarded their own name and symbol.

Before we look at the individual gates in more detail, let's briefly discuss the individual logic representations.

Schematic representation. This representation is useful when creating a detailed schematic diagram of a logic circuit and aids in pin assignments and debugging.

Truth table. A truth table lists every possible input combination and its resulting output. It provides a complete state-map of the component. By applying the truth tables to each circuit element and propagating the outputs to the next input you can (theoretically) always predict the behavior of any circuit. In reality, with large designs this approach can quickly become tedious.

Boolean logic. This representation is similar to algebraic notation although it has its own set of rules. Similar to algebraic expressions, Boolean logic expressions can be rearranged and simplified. When you work with them keep in mind that a fairly wide range of notations exists for the operators which can easily cause confusion. Also, be aware that some of the operators look similar to mathematical operators but they behave very differently in Boolean logic.

Verilog representation. Verilog is an officially sanctioned ANSI hardware descriptive language (HDL). Therefore, its notation has to be precise and unambiguous. It is very unforgiving and a typo can lead to unanticipated outcomes.

Venn diagrams. The Venn diagrams, sometimes also called Johnston diagrams, for a NOT, an OR and an AND gate are shown on the right-hand side of figure 1.1. The

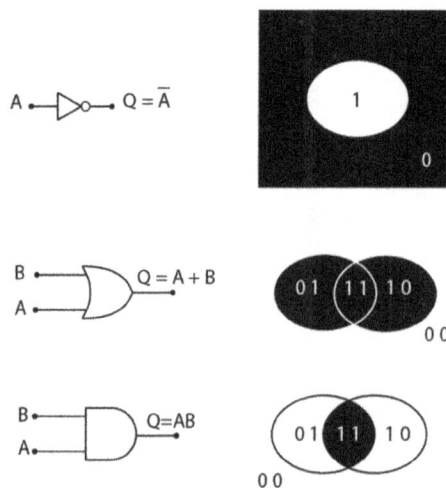

Figure 1.1. Venn or Johnston diagrams for a NOT, OR and an AND gate..

notation is based on set theory and can be used as yet another representation of digital logic. Similar to a truth table, it lists all possible input combinations with the first digit corresponding to the first input and the second digit to the second input. The dark shaded areas correspond to the regions resulting in a true output state for the particular logic operation shown. Venn diagrams are of limited use, in particular in larger circuits, and we will not use them any further.

1.2.1 Fundamental gates

The three fundamental gates of digital logic are the NOT, OR and AND gate.

The NOT gate is an inverter. It negates the current state and switches it to its opposite. For example, by NOT being 'on', something is 'off'. (*Note*: the NOT operator is sometimes written as $\bar{X}$, $\sim X$ or $!X$. As you will see later, if X represents a collection of bits then the meanings of $\sim X$ and $!X$ are not equivalent. Avoid the confusion and stick with $\sim X$.)

Applying a NOT operator successively an even number of times results in no change since every other NOT operation cancels the previous one. For example, applying the NOT operator twice as in: $Q = (\bar{\bar{X}})$ results in $Q = X$. A schematic representation of this expression is:

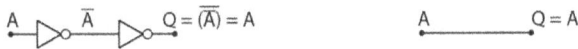

The NOT operator symbol is often drawn in an abbreviated form, omitting the triangle and using only the open circle at its end. The three circuits below illustrate this. All three circuits depict the functionality of a NOR gate, i.e. an OR gate followed by a NOT gate:

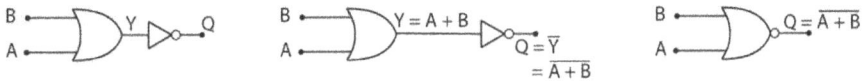

The same abbreviated form of a NOT gate is also used at the gate inputs, as depicted in the three circuits below. The circuits represent a 'NOT A, NOT B, OR-gate' (of course, the functionality of these three circuits is not equivalent to the three circuits shown directly above!):

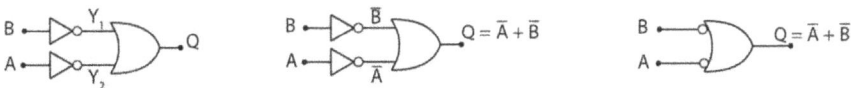

The remaining fundamental gates are the OR and the AND gates. Unlike the NOT gate which is a unary, single-input operator, both the OR and AND gates have at least two inputs.

First, note the commutative nature of these gates, and Boolean logic in general, shown below:

$$Q = A + B = B + A,$$

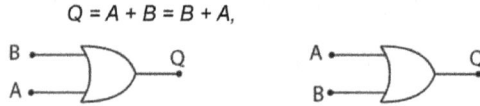

Although the gates shown above are two-input gates, we could easily expand their inputs to any number while maintaining their functionality. An example of a three-input OR gate is shown below. The circuit on the right is displayed in its concise form while the one on the left details its implementation with two-input OR gates. This approach could be extended further and it works for any two-input gate. Note that we implicitly made use of the Boolean associative properties, namely:

$$Q = (A + B) + C = A + B + C,$$

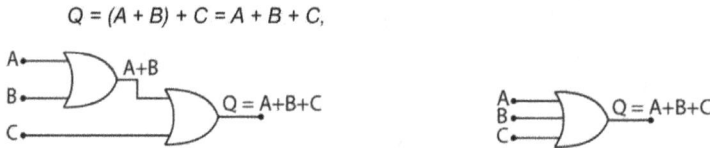

Finally, let's examine the difference between the OR and AND gates with the following two examples.

In the first example, assume that you have 1000 individual particle detectors and you want to monitor when at least one of them has been activated. Assume that when an individual detector is hit by a particle, the detector's output goes high. The easiest way to look for such an event is to connect all the particle detector outputs to one giant OR gate (having 1000 inputs) and monitor its (single) output. When the OR gate's output goes high, we know that at least one particle has passed through one of the detectors. In other words, the OR gate goes high when any one of its inputs goes high. This is the reason the OR operator is also known as the 'ANY' operator.

For the second example, consider that we are trying to locate a mysterious source of particles originating somewhere in the universe, far away. We will use a particle 'telescope' to detect their origin. It works similarly to an optical telescope, except that it employs a number of particle detectors in place of lenses, arranged in a straight line. We can pinpoint the particles' origin by sweeping our telescope across the sky and monitoring for an event when all our detectors are activated at the same time. This occurs only when all the detectors are aligned with the mysterious source. (In this example, we assume that the particles travel in a straight line and close to the speed of light.) This technique, often called an n-fold coincidence technique, where n refers to the number of detectors placed in coincidence, can be implemented by connecting all the detector outputs to the input of a giant AND gate. The output of the AND gate will go high only when ALL the inputs are high, meaning that we must have aligned our detectors in the direction of the source. This example illustrates the finicky nature of AND gates: unlike the 'combiner' OR gates, AND gates are very selective and act like a 'filter'. Since all its input conditions must be satisfied for its output to go high, an AND gate is sometimes called an 'ALL' gate.

Finally, a quick word on Boolean notation for these two gates: the OR operator is usually expressed by using the addition symbol, +. However, Verilog does have an addition operator and this can lead to confusion since the behavior is different. This will be discussed in more detail later. In Boolean notation, the AND operator is often omitted and it is implicitly understood that AB really means $A \cdot B$.

1.3 Basic Boolean logic rules: application and examples

Example 1: timing diagram

The diagram in figure 1.2 shows the output from various gates based on the time-dependent input of A and B. In this timing diagram the x-axis represents time and the y-axis the digital voltage level. (*Note*: the last trace shows the output from an XOR gate.)

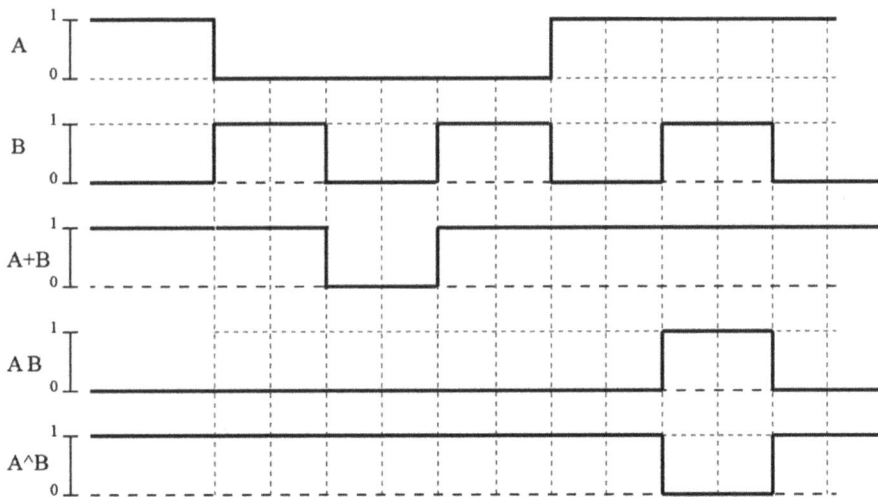

Figure 1.2. Time dependent output for an OR, AND and XOR gate as a result of input signals A and B.

Example 2: deriving basic Boolean logic rules from truth tables

Let's apply the truth tables from table 1.2 to derive some fundamental Boolean logic rules. See if you can work out the conclusion from the expression on the left by using the truth tables for the particular gate.

Table 1.3. This table lists some basic Boolean expressions and their corresponding schematic representation. It utilizes truth tables to obtain the resulting state.

Expression	Schematic symbol	Truth table			Conclusion
$Q = A + 0$		A	B	Q	$A + 0 = A$
		1	0	1	
		0	0	0	
$Q = A + 1$		A	B	Q	$A + 1 = 1$
		1	1	1	
		0	1	1	
$Q = A \cdot 0$		A	B	Q	$A \cdot 0 = 0$
		1	0	0	
		0	0	0	
$Q = A \cdot 0$		A	B	Q	$A \cdot 1 = A$
		1	1	1	
		0	1	0	
$Q = A + \bar{A}$	Alternative schematic:	A	B	Q	$A + \bar{A} = 1$
		1	0	1	
		0	1	1	
$Q = A \cdot \bar{A}$		A	B	Q	$A \cdot \bar{A} = 0$
		1	0	0	
		0	1	0	
$Q = A + A + A$ $= ((A + A) + A)$		A	$A+A$	Q	$A + A + A = A$
		1	1	1	
		0	0	0	

1.4 Interchangeability of gates and De Morgan's theorem

In the previous exercise, we simplified Boolean expressions and eliminated redundant components. This section will use the truth tables and show that we can mimic the behavior of one type of gate by substituting it with a different type or types. In other words, we will develop Boolean rules on how to interchange gates.

Example 1. NOT gate from a NAND gate

Shown below is a NAND gate with its inputs tied together. When you examine its truth table (below), you notice that it functions as an inverter! Note that the same result could have been obtained by using a NOR gate in a similar configuration.

A	$B = A$	AA	$Q = \overline{AA}$
1	1	1	0
0	0	0	1

Conclusion: you can substitute a NOT gate with a NAND or NOR gate whose inputs have been tied together. However, you can never create a NOT gate from an AND or an OR gate alone. (Proof left to the reader.)

Example 2. AND gate from NAND gate

In this example, we negate the output of the NAND gate with another NAND gate disguised as a NOT gate. The double negation of the AND gate output results in its original AND gate. The conclusion is that NAND or NOR gates can be substituted for NOT, OR or AND gates. Similar to the previous conclusion, OR and AND gates alone cannot be substituted for NAND or NOR gates.

Example 3. De Morgan's theorem

This last example is so important to Boolean logic that it has even been given its own name, De Morgan's theorem. Examine the truth table for the circuit shown below. What type of gate does its output Q remind you of?

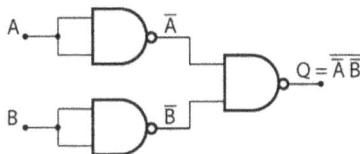

Applying a truth table to the circuit above yields the truth table:

A	B	$\bar{A}$	$\bar{B}$	$\bar{A}\,\bar{B}$	$Q = \overline{\bar{A}\,\bar{B}}$
0	0	1	1	1	0
0	1	1	0	0	1
1	0	0	1	0	1
1	1	0	0	0	1

Although this circuit only uses NAND gates, its output, Q, behaves like an OR gate. By symmetry, it follows that we should also be able to use a combination of NOR gates to act like an AND gate. This is indeed the case and its circuit is shown below. Prove to yourself using a truth table that for the circuit below $Q = AB$:

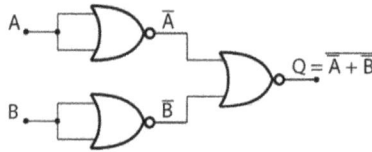

The implication of these two examples is profound: they show that we can implement any circuit or Boolean expression by using only one type of gate. In other words, all digital logic can be implemented with a single type of gate, that is, as long as it uses only NAND or NOR gates!

In more formal terms, the two De Morgan's theorems are:

$$\overline{AB} = \bar{A} + \bar{B}$$

$$\overline{A + B} = \bar{A}\bar{B}.$$

(*Note*: these two equations represent the general form of De Morgan's theorem and not the circuit shown directly above them.)

A simple way to remember the theorem is to notice that when you 'cut' the negation bar between two terms you must always change the operator from an AND to an OR, or vice versa.

Applying De Morgan's theorem to the first circuit diagram shown above, we can now prove that it represents an OR gate:

$$Q = \overline{\bar{A}\bar{B}} = \bar{\bar{A}} + \bar{\bar{B}} = A + B.$$

Similarly, the second circuit using the NOR gates becomes:

$$Q = \overline{\bar{A} + \bar{B}} = \bar{\bar{A}}\bar{\bar{B}} = AB$$

De Morgan's theorem should remind you that extending or cutting the negation operator without changing the operator below it leads to an unequal expression. This very common mistake is as follows:

$$\bar{A}\bar{B} \neq \overline{AB}.$$

1.5 Implementing digital logic

1.5.1 Implementing truth tables with Boolean logic expressions

In the preceding sections we derived and manipulated Boolean expressions using truth tables. In this section we convert truth tables to their corresponding Boolean logic expressions.

A formal procedure for implementing truth tables involves Karnaugh maps, detailed information can be found online. However, for our purpose, a simplified version of such a procedure is provided by the algorithm listed below. It will work with any arbitrary truth table and will produce a Boolean logic expression that correctly represents the original table. However, unlike other algorithms, it will not necessarily provide the most optimized expression. This is not really an issue for us since our applications are simple and, if needed, the Verilog compiler will optimize the expressions for us.

Example 1. Finding the Boolean expression for an XOR gate
Attesting to its usefulness, the XOR gate has its own schematic symbol and operator. However, it is not a fundamental gate and it can be constructed with AND, OR and NOT gates. We will now show how to derive its Boolean expression starting with its truth table:

A	B	Q
0	0	0
0	1	1
1	0	1
1	1	0

Although we may be able to figure out a logic expression by trial and error, it can be implemented more systematically by noting the following observations:

States	A	B	Q_i	Q_i
0	0	0	0	Q_0
1	0	1	1	Q_1
2	1	0	1	Q_2
3	1	1	0	Q_3

Step 1. Note that the *output* of the entire truth table is represented by the four states labeled 0 to 3 in the table above. In other words, we could express the table with the following general Boolean expression involving four OR operations:

$$Q = Q_0 + Q_1 + Q_2 + Q_3, \tag{1.1}$$

where

$$Q_0 = 0, \; Q_1 = 1, \; Q_2 = 1 \text{ and } Q_3 = 0. \tag{1.1a}$$

Step 2. We have already shown that $X + 0 = X$. Therefore, we can eliminate all states with $Q_i = 0$ since these will not affect our result. In our example, we end up with following simplified expression of Q:

$$Q = Q_1 + Q_2. \tag{1.2}$$

In other words, Q_0 and Q_3 have been eliminated since they are 0.

Step 3. In each state, Q_i is a function of its inputs, i.e. $Q_i = Q_i(A_i, B_i)$. For our case this means

$$Q = Q_1(A_1, B_1) + Q_2(A_2, B_2). \tag{1.3}$$

Now we have to find a Boolean expression for the case listed that will result in a high output for its, and only its, specific input. Therefore, we must find a Boolean expression that satisfies the following condition:

$$Q_1(A_1, B_1) = Q_1(A = 0, B = 1) = 1, \tag{1.4}$$

$$Q_2(A_2, B_2) = Q_2(A = 1, B = 0) = 1. \tag{1.5}$$

The only expressions satisfying our requirement for Q_1 and Q_2 are

$$Q_1 = \bar{A}B, \tag{1.6}$$

$$Q_2 = A\bar{B}. \tag{1.7}$$

(It should come as no surprise that such a specific request can only be satisfied with an AND operator.)

Step 4. Putting equations (1.2), (1.6) and (1.7) together results in

$$Q = Q_1 + Q_2 = \bar{A}B + A\bar{B}. \tag{1.8}$$

This is the Boolean definition of an **XOR** gate and we have successfully implemented the truth table.

Example 2. Three-input truth table to Boolean expression

Simple two-input truth tables can often be evaluated with a trial, error and good luck approach. However, more complex truth tables may resist this approach and may require the algorithm shown above. For example, consider a three-input truth table which means that Q is now composed of eight states:

$$Q = Q_0 + Q_1 + Q_2 + Q_3 + Q_4 + Q_5 + Q_6 + Q_7.$$

The rest follows the outline above. See if you can implement the table below and show that it conforms to the Boolean expression listed below the table:

States	A	B	C	Q	Qi
0	0	0	0	0	Q_0
1	0	0	1	1	Q_1
2	0	1	0	1	Q_2
3	0	1	1	0	Q_3
4	1	0	0	1	Q_4
5	1	0	1	0	Q_5
6	1	1	0	0	Q_6
7	1	1	1	1	Q_7

$$Q = \bar{A}\bar{B}C + \bar{A}B\bar{C} + A\bar{B}\bar{C} + ABC.$$

1.5.2 Creating a schematic representation from a Boolean expression

Now that we know how to derive a Boolean expression from a truth table, let's see how to represent it with a digital circuit. Start by working your way from the most general expression at the output back to the input and identify its basic operators. For example, starting at the output, the expression derived at the end of the previous section can be rewritten as

$$Q = Q_1 + Q_2 + Q_4 + Q_7.$$

Each of the four terms, Q_1, Q_2, Q_4 and Q_5, are joined by an OR operation which can be represented with the schematic symbol of a four-input OR gate.

Working our way to the inputs, the terms Q_i are composed of three-input AND gates which may contain negation operators at some of their inputs. (Remember: the Boolean symbol for the AND operation, a middle dot ' · ', is usually dropped and, hence, often overlooked.)

$$Q_1 = \bar{A}\bar{B}C,$$

$$Q_2 = \bar{A}B\bar{C}$$

$$Q_4 = A\bar{B}\bar{C},$$

$$Q_7 = ABC.$$

Putting this all together results in the schematic shown below:

We have made it easy on ourselves by using multi-input gates, such as the three-input AND and the four-input OR gate. If we were forced to use only two-input gates we could replace these gates with a series of two-input gates as shown earlier.

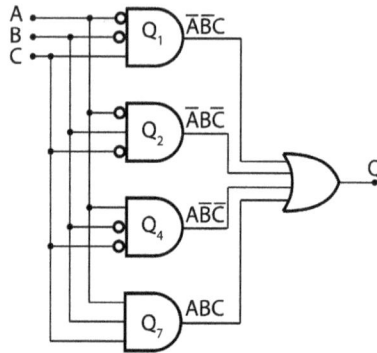

1.5.3 Creating a Boolean expression from a schematic representation

To convert a schematic representation to a Boolean expression, start at the inputs and then work through each gate, inserting after each gate the corresponding Boolean expression. (This is the opposite order of the previous example.) See the example above where we have omitted the final Boolean expression for the circuit since it has already been stated in the text.

1.6 Tri-state logic

Carefully study the circuit shown below. What is its output Q when $A = 1$, $B = 0$ and $C = 0$? Since $A + B = 1$ and $B + C = 0$, will Q be high or low? The answer is that Q's state cannot be predicted because two outputs in opposing states are connected together. This arrangement shorts the outputs of the two OR gates forcing them to fight each other whenever they are not equal. Depending on the physical design of the OR gates, one output will probably prevail over the other, although it is difficult to know which one that will be. (Additionally, situations like these often draw large currents and can create current and voltage spikes which will mess up the logic states of the gates in the circuits.)

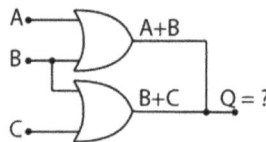

Therefore, you must never connect the output of one gate to the output of another gate! Remember that outputs can only be connected to another input. It follows then that the circuit above is meaningless.

Having made such a strong case against ever connecting outputs together, in a few rare instances we are simply left with no choice. Such situations can arise in communication circuits when the same wire is used to talk and listen to a device, a situation referred to as a 'bidirectional wire'. In such a case it is unavoidable that multiple outputs are connected. However, at any one time, only one device's output can be active and all the others must be disabled.

The digital technology that allows this kind of behavior uses tri-state gates. While the familiar two-state gates can only be in an active-low or an active-high state, tri-state gates have an additional third state, typically called a HIGH-Z state. (Another expression might be a 'disabled,' or 'don't care' state.) When the tri-state gate is in this state, its output is physically disconnected from the circuit through an infinitely large output resistance, hence its name HIGH-Z.

It works by attaching to every gate's output, a switch as shown in the diagram below. When the gate is in the HIGH-Z state, i.e. the switch is open, it no longer affects or is affected by the circuit connected to it.

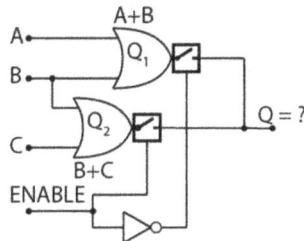

To set or reset the HIGH-Z state, each gate requires an additional input, labeled ENABLE. As we have already indicated, for this type of circuit to work, at most only one of the gate's outputs can be in the active state while all the others gates remain in the HIGH-Z state. The two-gate circuit shown above accomplishes this by using an inverter to ensure that only one of the two gates is active at any given time.

While the tri-state logic is important for large digital designs, we will only briefly touch on it in our exercises and almost all of our circuits will be using the two-state logic discussed so far.

1.7 Boolean logic expression summary

Table 1.4 shows a summary of the Boolean logic expressions we have discussed. You are now ready to test your knowledge of digital logic with the exercises in the following chapters.

Table 1.4. Summary of relevant Boolean expressions.

$A + 0 = A$

$A\,0 = 0$

$A + 1 = 1$

$A\,1 = A$

$A = A + A = A + A + A$

$A + \bar{A} = 1$

$A\bar{A} = 0$

$\bar{\bar{A}} = A$

$A + B = B + A$

$AB = BA$

$A + (B + C) = (A + B) + C$

$A \oplus B = A\bar{B} + \bar{A}B$

$\overline{AB} = \bar{A} + \bar{B}$

$\overline{A + B} = \bar{A}\bar{B}$

$AB + A\bar{B} = A(B + \bar{B}) = A\,1 = A$

1.8 Exercises

Challenge yourself and see if you can prove the following five expressions:

1. $A + AB = A$.
2. $A\bar{B} + B = A + B$.
3. $A + \bar{A}B = A + B$.
4. $AB + BC + \bar{A}C = AB + \bar{A}C$.
5. $\overline{A \oplus B} = AB + \bar{A}\bar{B}$.

Chapter 2

FPGA and VERILOG: combinational logic I

Additional reading

This chapter is (mostly) self-contained and very long. It contains a large amount of detailed information about digital logic and the Verilog language. You should definitely read it before proceeding further.

Briefly read sections 11.1 and 11.2 (pp 764–70) in Horowitz P and Hill W 2015 *The Art of Electronics* 3rd edn (New York: Cambridge University Press).

A detailed reference manual of the BASYS board (including pinouts) can be found at Digilent's website: https://reference.digilentinc.com/_media/basys3:basys3_rm.pdf.

2.1 Introduction to digital hardware

In this and the following three chapters you will explore digital logic concepts and the hardware associated with it.

In the very first exercise, you will use discrete logic gates and explore their properties. While it is (still) possible to purchase and use these integrated circuits, most contemporary systems have replaced these 'fixed logic' chips (also known as application-specific standard parts (ASSPs or ASSP devices) with programmable logic gates using, for example, field programmable gate arrays (FPGAs) or complex programmable logic devices (CPLDs).

The demise of the discrete ASSP chips is due to the increasing complexity and speed that most contemporary digital projects require. Single Boolean gates are extremely basic. A useful or interesting design requires a large number of these basic gates. Interconnecting these ASSP chips with wires creates delays and crosstalk, limiting these designs to a maximum speed of a few megahertz.

On the other hand, an FPGA, like the one you will be using, contains a million programmable gates and operates reliably up to hundreds of megahertz. While the FPGAs solve most (of our) design requirements, they come with some overhead: first, you will have to program the FPGA with your design before you can use it. Second, typical FPGA chips have anywhere from 100–300 pins, which makes it difficult, or rather impossible, to connect them to a bread board, like the ones typically used for prototyping electronic circuits. As you can see, you will need some additional tools to use the FPGAs.

First, you will need a hardware descriptive (programming) language (HDL) to specify your digital design. In this book we will be using a language called **Verilog**. You will learn it as you work through the exercises in the next few chapters. You will also need a software or compiler to synthesize your Verilog designs so that you can program the FPGA. You will be using Xilinx **Vivado's** integrated development environment (IDE). This software has to be installed on your computer and you may download it (for free). However, one warning: it requires close to 10 GB of disk space.(Installation instructions are given in appendix A).

Finally, you will need the actual FPGA hardware. You will use a programmable logic board produced by Digilent Inc., a **BASYS3** board. In addition to the Artix-7 FPGA from Xilinx, the board also comes with connectors, displays, switches, LEDs and the support machinery to let you program the FPGA from your computer through the USB port. You can find a link to the full documentation for BASYS3 in the appendix at the end of this book.

To use an analogy, you may view the FPGA as a blank digital canvas on which you will paint your digital design using Verilog and Vivado. So let us get started!

2.2 Application specific standard parts (ASSP): 74XX digital logic chips

Before FPGAs became ubiquitous, digital logic circuits were often implemented using the 74XX family of ASSP logic gates. These 14-pin chips usually contain multiple gates of the same type, like the 74X86 shown in figure 2.1, which comes

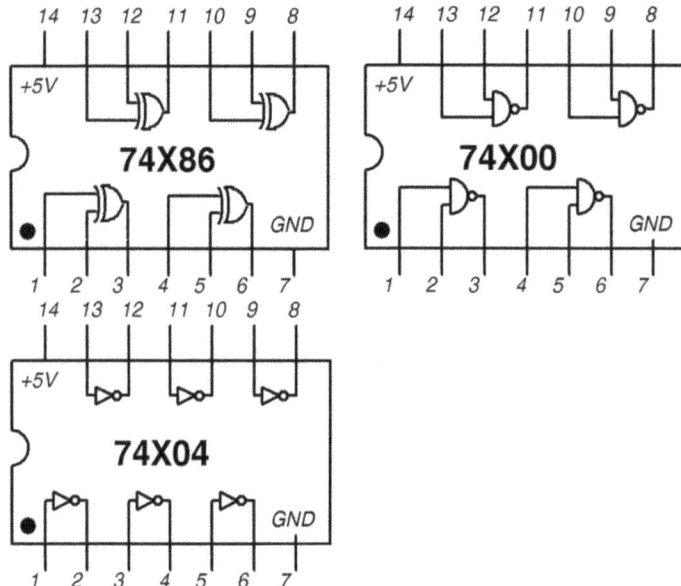

Figure 2.1. Pin-outs of the 74X86 XOR, 74X00 NAND and the 74X04 NOT logic gates.

with four individual XOR gates. (*Note*: X stands for any letter and designates a specific subcategory, for example, low power consumption or high speed).

These chips adhere to the traditional 0 and +5 V TTL logic standard and you must always supply these voltages to pins 7 and 14, respectively. The chips will not work without these supply voltages.

The output of an XOR is high only when the two inputs to the XOR gate are low and high, or high and low. This can be expressed as

$$A \oplus B = \bar{A}B + A\bar{B}.$$

Work out the complete truth table for the XOR gate and test it with an actual 74X86 chip. Power it as shown above and select one of the four XOR gates and set its inputs either high (+5 V) or low (ground). Instead of observing the gate's output on the scope, connect it through a 330 Ohm series resistor and an LED to ground:

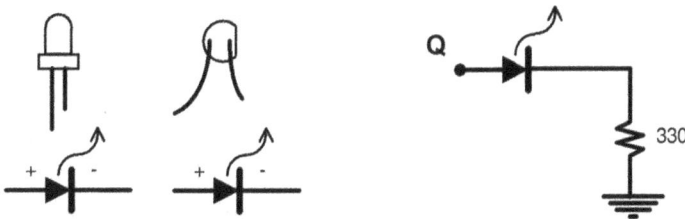

Next, check what happens if one input is left 'floating', i.e. leave one of the gate's inputs unconnected while you set the other to high and then to low. From your results, is the floating input considered low or high?

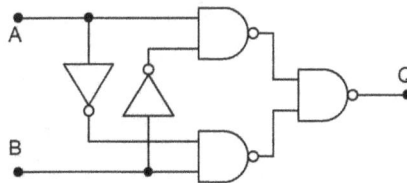

Convince yourself that the Boolean circuit shown above is an implementation of an XOR gate based on the previous equation with some of the gates substituted with their NAND equivalents. Construct it using a 74LS00 (quad NAND) and a 74LS04 (hex NOT) chips and try it out. Again, connect a series resistor and an LED to the output and confirm that it behaves identically to the previous 74X86 XOR gate. (If you are bored or need an extra challenge, see if you can construct the XOR circuit using only 4-NAND gates.)

2.2.1 Review exercises

1. Show your truth table for the XOR expression and the actual 74X86 XOR gate.

2. Are the unused or floating inputs considered to be low or high for this logic family? Explain how you arrived at this answer.
3. Show the truth table you obtained with the XOR gate constructed from the NAND and NOT gates.

2.3 Introduction to the Vivado design environment and the BASYS3 boards

2.3.1 Objective

In this first FPGA exercise you will familiarize yourself with Digilent's BASYS3 board and the Vivado HL Webpack integrated design environment.

The goal of your first project is rather modest. You will program the FPGA to invert the state of the LEDs, which in turn are controlled by the switches. In other words, the LEDs will turn off when switches are on or in the upward position. (This is the opposite

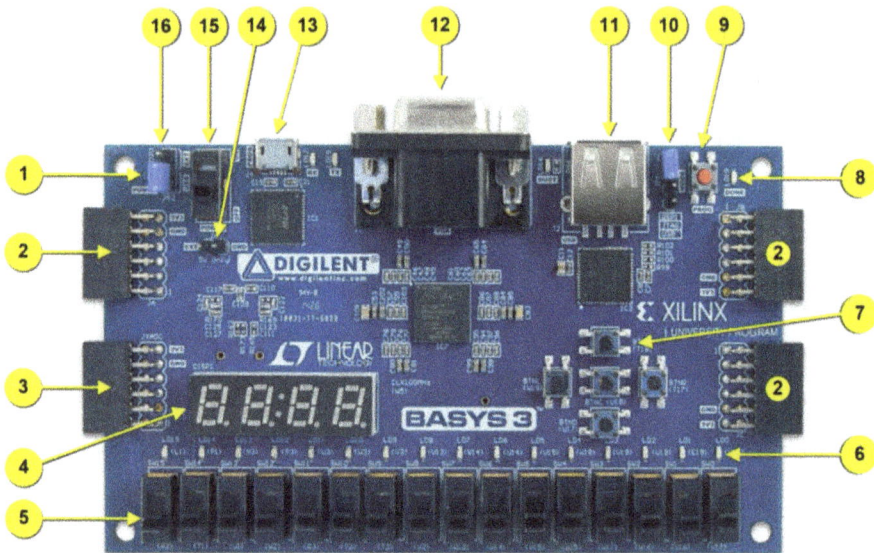

Basys3 FPGA board with callouts.

Callout	Component Description	Callout	Component Description
1	Power good LED	9	FPGA configuration reset button
2	Pmod connector(s)	10	Programming mode jumper
3	Analog signal Pmod connector (XADC)	11	USB host connector
4	Four digit 7-segment display	12	VGA connector
5	Slide switches (16)	13	Shared UART/ JTAG USB port
6	LEDs (16)	14	External power connector
7	Pushbuttons (5)	15	Power Switch
8	FPGA programming done LED	16	Power Select Jumper

Table 1. Basys3 Callouts and component descriptions.

Figure 2.2. BASYS3 board with a Xilinx's Artix 7 XC7A35T-1CPG236C FPGA in the center. (Reproduced with permission from *Basys3™ FPGA Board Reference Manual* (2016, Pullman, WA: Digilent). Copyright 2016 Digilent Inc.).

of the default state when you turn the board on.) We chose this simple exercise so that you are able to acquaint yourself with the BASYS3 board, the programming environment of Vivado, and the HDL programming language itself, Verilog.

2.3.2 Hardware: BASYS3 board with the Xilinx Artix 7 FPGA

Connect your BASYS3 board through the micro USB port (callout 13 in figure 2.2) to your computer. Make sure the On/Off switch (15 in figure 2.2) has been turned on. To confirm that your board is on and works, LEDs on callouts 1 and 8 should be lit.

2.3.3 Vivado project development: overview

In the steps outlined below, you will learn how to create and implement a complete Vivado project. You will start with a completely new project and then add the necessary code in Verilog to control the FPGA. Once you have compiled or generated the bit-file, you will use it to program the board.

Step 1. Creating a new project

Start by opening the latest version of Xilinx Vivado: 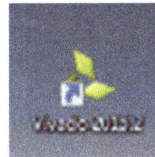 . The first screen you will see is shown below.

Click on the 'Create New Project' icon.

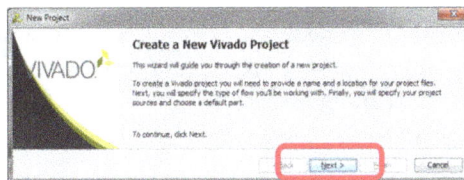

After the message window (shown above) appears you are asked where you want to store your project. Due to some dubious restriction within Vivado, you are *not* allowed to create the project in a folder that is linked through a UNC (universal naming convention), such as for example 'U:\\spa-home\mxpusers'. So, if your computer is connected to a network drive and you were to select a folder within your 'My Documents' folder, Vivado will prevent you from proceeding.

Therefore, you are forced to store your project in a mapped folder, such as the 'C:\YourX500Name\My Documents' or your personal USB drive. You may find that storing your projects on your personal hard drive (as opposed to an USB drive) considerably speeds up Vivado's compile time.

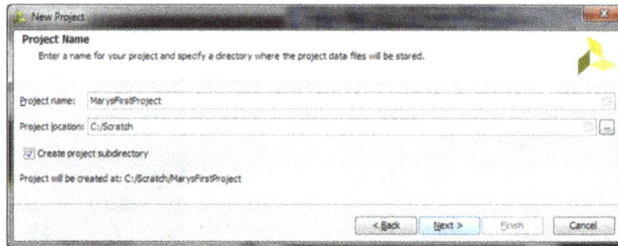

Select the default resistor–transistor-logic ('RTL') project and make sure that the 'Do not specify sources at this time' box is checked. Click 'Next'.

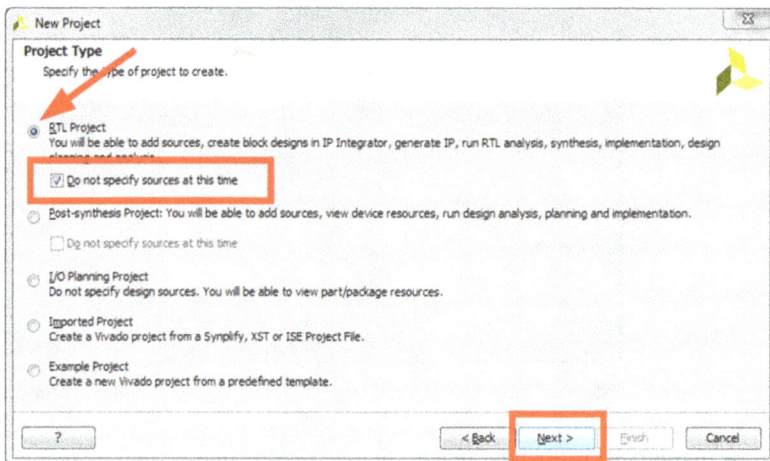

If you should have forgotten to check the 'Do not specify sources…' box, you will get the screens below asking you for source files. For now you may ignore them and simply skip through the next three screens.

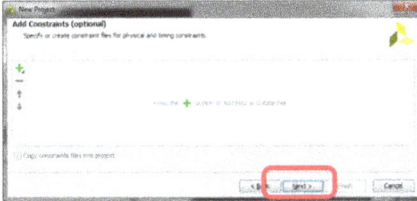

At this point you must specify the detailed hardware specifications of the FPGA that you will be using, such as its temperature and speed rating.

If you followed the instructions in the appendix and have already installed the BASYS3 board support files your life will be a bit easier: all you have to do is to click on the 'Boards' tab and then select the appropriate board, i.e. BASYS3, and Vivado will link it to the correct FPGA hardware specifications.

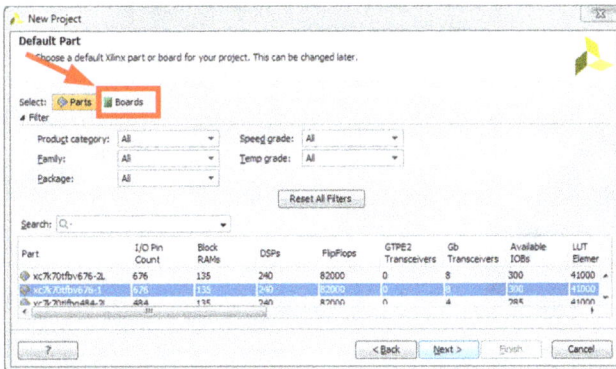

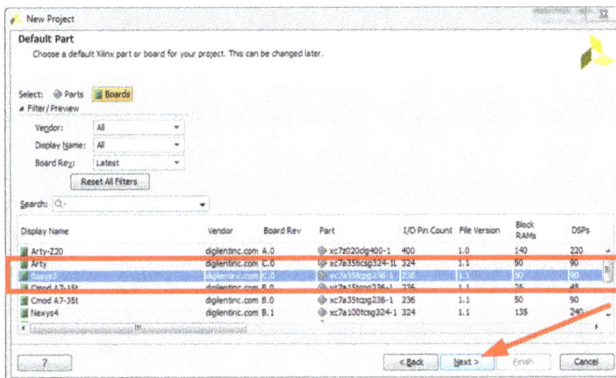

However, if you are working on a computer where the board files have not been installed you need to select *all* the correct FPGA hardware specifications yourself, as shown in the window below.

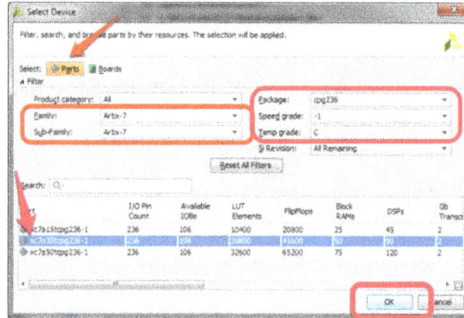

In either case, your final 'New Project Summary' (shown below) must list the correct FPGA, namely an 'Artix-7', 'xc7a35t' in a 'cpg236' pin package with a speed grade of −1. (Yes, the FPGA chip has 236 pins! Are you not relieved that you do not have to wire them all up?)

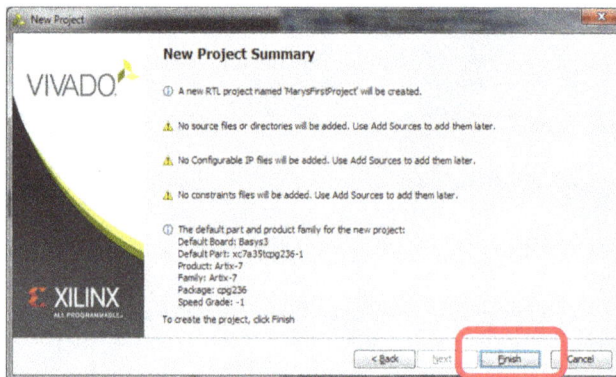

Your pop-up message may look slightly different depending whether or not you added any sources.

Summary: So far, you have created a new project for a specific FPGA. However, you have not yet specified what the FPGA will be implementing. This step is done in the next section.

Step 2. Creating, adding and modifying the Verilog file

Before we add the Verilog source file to the project, spend a moment familiarizing yourself with Vivado's development environment, shown below.

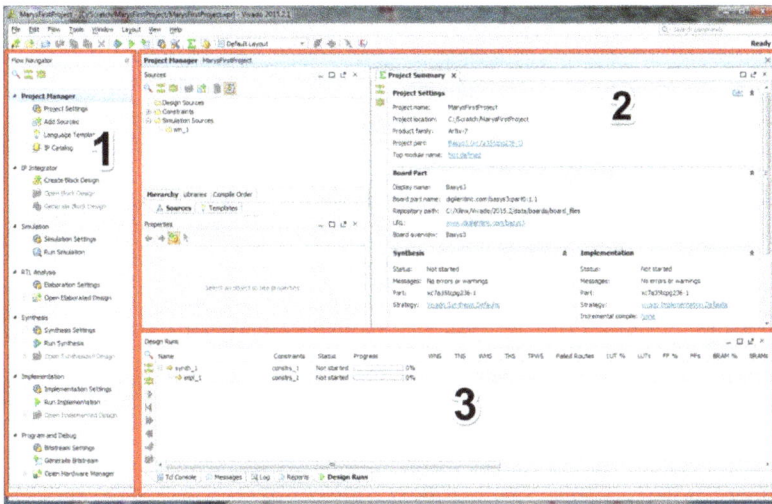

Aside from the menu bar on top, it consists of three main sections. In the left-most panel, panel 1, the 'Flow Navigator' lists in sequential order all the steps required for creating the bitstream data file to program the FPGA. You may at any time move up or down the hierarchy and depending on your choice the program will complete any omitted steps. The top right panel, panel 2, lists the files in the project. We have not yet added any files, so it is still empty. Also, note that the content of panel 2 depends on the specific step you have selected in the 'Flow Navigator', panel 1. The bottom panel, panel 3, displays status messages and the progress of the compile process.

Now let us add a *design source file* to the project. Its purpose is to specify the FPGA's tasks. Click on the 'Add Sources' label in the 'Flow Navigator/Project Manager' (see below).

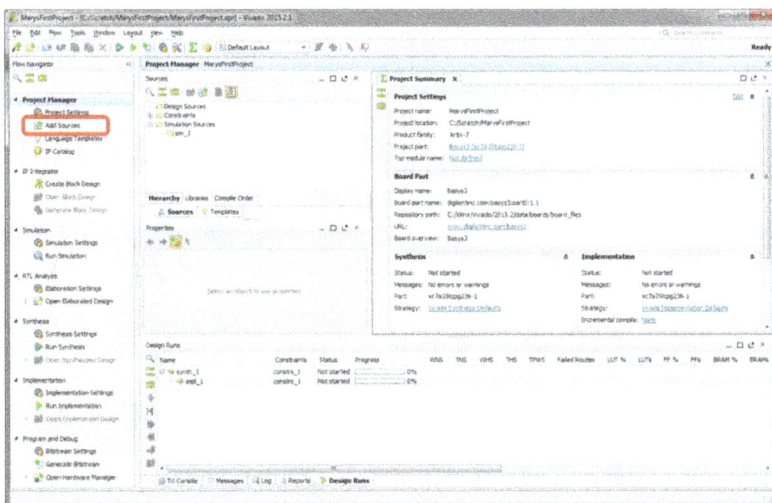

The following window opens.

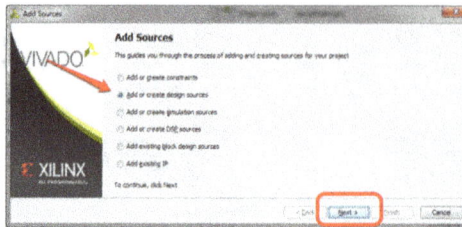

Next, you may either click on the 'plus sign' and then select 'Create New File...' or you may directly click on 'Create File'.

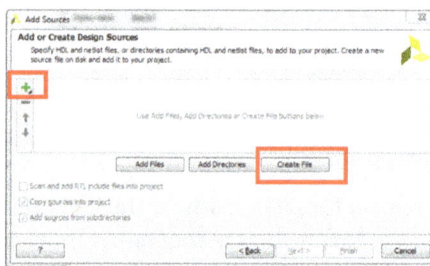

The window below opens asking you specify a (unique) name for your design source file. (At this point note that the default language or 'File Type' has been selected as 'Verilog'. This is one of the two most common HDLs for these types of files, the other being 'VHDL'. Throughout the rest of this book we will use Verilog.) Name your file and hit 'OK'.

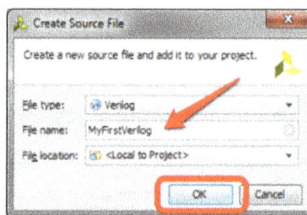

A confirmation window opens up. Click on 'Finish'.

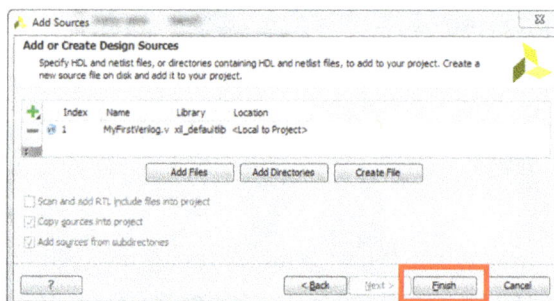

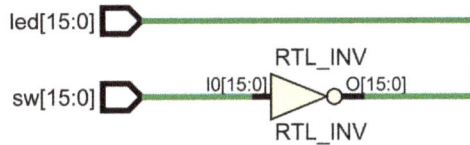

Figure 2.3. The elaborated design schematic of our first project: each of the 16 switch inputs, 'sw', is negated and then sent to its corresponding LED output, 'led'.

After clicking the 'Finish' button, Vivado is ready to create the framework of your Verilog code. Specifically, this entails specifying and naming the input/output (I/O) ports for your design.

Shown in figure 2.3 is the schematic design for the project that we want to implement. Although you are (theoretically) free to select any suitable name for the I/O ports, we have deliberately selected the name 'sw' for the switches' input port and 'led' for the LEDs' output port. (You will better understand why we have selected these names when you get to the next section where we map the ports to the physical pins of the FPGA.)

Note the use of the bracket notation after the I/O port names, such as [15:0]. This specifies the size, or width, (in bits) of the port. In other words, in our example, all ports consist of 16 (single-bit) wires labeled from 0, the least significant bit, to 15, the most significant bit. The notation is similar to a column vector in mathematics and that is why such multi-bit ports are usually called 'vectors' or also buses.

Important: also note that the schematic design above only shows one inverter. That could be misleading but note that its input and output are vectors, thereby indicating that there are really 16 individual inverters in our design, each connected to its corresponding switch input and LED output.

Now let us enter all this information in the screen's I/O port definition window, as shown below. Be sure to define the correct direction of the ports, i.e. are they inputs or outputs?

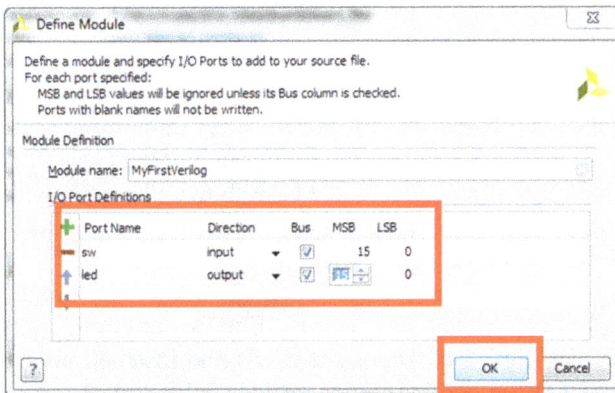

As already stated, stick with the port names shown above.

After you have entered the port information and pressed the 'OK' button, Vivado will generate the Verilog file and then return to its development environment (see below). To provide you with your first glance of Verilog code, open and examine the Verilog file you just have created. With the 'Project Manager' selected, in the 'Sources' window, double-click directly below the 'Design Sources' on your Verilog file (see step 1 below). Your newly generated Verilog file opens up in panel 2.

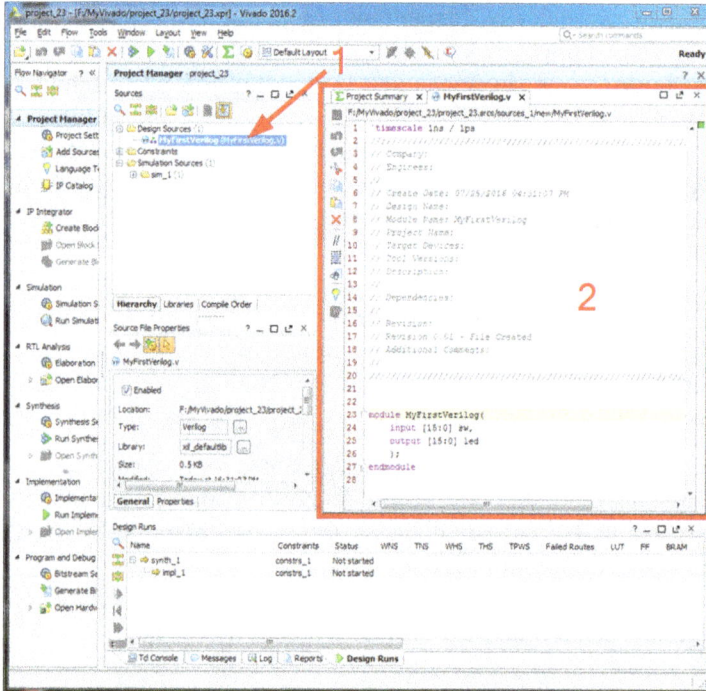

Note the color coding of the Verilog statements. Most of the lines in the newly created Verilog file are grayed out to indicate that they are comment statements. These lines are optional and they provide additional information and can be used to enhance the readability of the code. In the case shown above they take up an entire line but comments can also be appended to a Verilog statement as long as they are preceded by a double backslash, //.

All the words displayed in purple, such as 'module', 'input', 'output' and 'endmodule', represent Verilog key words. Finally, anything in black, such as 'MyFirstVerilog', 'sw' and 'led', represent labels and instructions we have entered.

The core of the Verilog code you created is listed below. It represents the most fundamental unit of the Verilog language, namely a *module*:

```
module MyFirstVerilog(
    input [15:0] sw,
    output [15:0] led
    );

endmodule
```

Similar to a mathematical function, every Verilog module has a name; in the above case the module is named 'MyFirstVerilog'. Instead of arguments, a Verilog module has I/O ports, such as 'sw' and 'led'. However, unlike a mathematical function where the arguments are always 'inputs', a Verilog module's I/O ports can either be an input or an output. Therefore, its direction must be explicitly specified with the Verilog keyword 'input' or 'output'.

Again, notice the use of vectors. It is a convenient way to group a number of similar ports such as the individual switches or LEDs. In the sample code shown above, there are 16 (1-bit) input ports, namely sw[0], sw[1], sw[2] through sw[15], which were grouped into a 16-bit vector named sw[15:0]. In a similar fashion, the output ports led[0], led[1] through led[15] were grouped into vector led[15:0]. Just like with matrix algebra, we will later see that grouping individual ports into vectors allows us to perform powerful operations on a large group of bits.

At this point the Verilog code shown represents a mere framework of a module; it only defines its I/O ports and their respective directions. What the Verilog code (yet) lacks are specific instructions for the signals applied to these ports. It is your job to add this information to the code created. Specifically, since the goal of this exercise is to connect all the switches to the LEDs (we will apply the inverting in the next paragraph), specify this using the Verilog 'assign' keyword as shown below:

```
assign led = sw;
```

You may regard the 'assign' keyword as a command to directly connect the port(s) on the left-hand side of the equal sign to the expression on the right-hand side. In the case of a vector, the assign statement wires each bit on the right-hand side to its corresponding bit on the left-hand side. For example, in the above case, led[0] will be wired to sw[0], led[1] to sw[1,] etc. (You will explore what happens in an assignment statement if the two vectors have different sizes in the next section.)

Since the purpose of this exercise is to turn the LEDs off when the switches are on and vice versa, you will make use of the Verilog bit-wise negate operator, ~. Add it to your assign statement as shown below:

```
assign led = ~sw;
```

What we have said before still holds true: each bit on the left-hand side will be wired to the corresponding bit expression on the right, i.e. led[0] will be wired to its corresponding expression, i.e. ~sw[0], led[1] to ~sw[1], etc.

(We will have more to say about bit-wise Verilog operators in the next section.)

Let us complete the Verilog module by *adding* this statement to your Verilog source code as shown below in box 1.

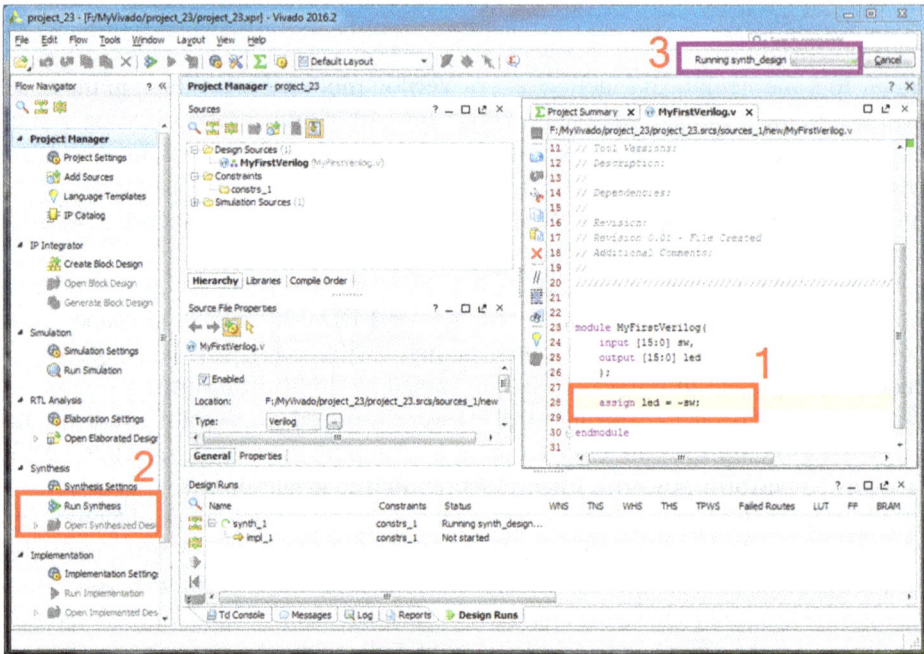

In the 'Flow Navigator', click 'Synthesis/Run Synthesis', box 2 in the picture above. The synthesis checks the syntax of your Verilog code for syntax errors. You can monitor its progress in the window in the upper right corner, labeled 3. Once it completes successfully, the window below opens. For now we will select 'Cancel'.

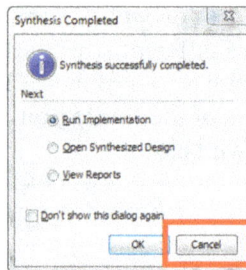

To sum up, you have now created a complete Verilog program that specifies the I/O ports and what happens to the signals applied to them. What we still are lacking is a way to connect the functionality of the Verilog code to the physical hardware of the BASYS3 board. We will remedy this in the next section where we will map specific pins of the FPGA with the I/O ports specified above.

Step 3. Creating a constraint file to map the Verilog I/O ports to the FGPA pins

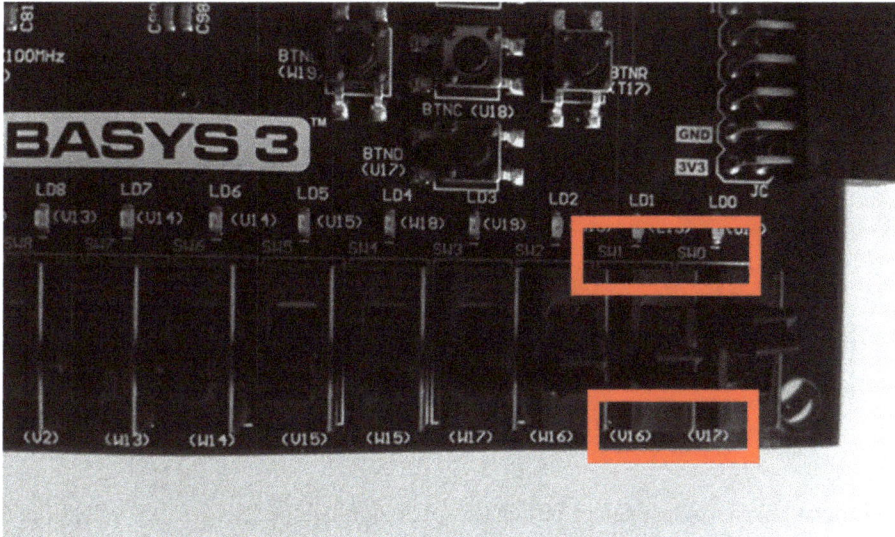

Figure 2.4. Close up of the BASYS3 board. The component names for switch 0 and switch 1, 'SW1' and 'SW0', have been highlighted as well as their pin assignments, 'V16' and 'V17'.

When you look at the picture of the BASYS3 board shown in figure 2.4, you will notice two types of labels printed next to the components. First, the labels not enclosed by parentheses, such as SW0 and LD0, identify the board components such as switch 0 and LED 0, respectively. The labels adjacent to these and enclosed in parentheses, such as (V17) and (U16), specify the pins of the FPGA to which these components have been physically connected by the board manufacturer.

In our Verilog module, we used ports 'sw' and 'led' and we now want these to be mapped to the physical switches and LEDs. For example, we want sw[0] to be mapped to the physical switch SW0 at pin V17; similarly we want sw[1] to be mapped to SW1 (at pin V16) etc, up to and including port sw[15] at SW15 at pin R2. To achieve this we must explicitly declare these pin-to-port mappings in a (universal) *constraint file* which will then be added to the project.

As you can imagine, creating such a constraint file from scratch can be very tedious. You will need to list all the ports and then look up the corresponding pin numbers and specify the corresponding voltages. Instead we will employ an already existing template that lists all the BASYS3 ports and pin numbers. We will add this file, called the 'BASYS3_Master' constraint file, to the project and then modify it to suit our project. If you have not done so already, download the BASYS3 master constraint file from https://github.com/Digilent/Basys3/tree/master/Resources/XDC.

Let us add the master constraint file. In the 'Flow Navigator' select 'Project Manager/Add Sources' (point 1 below) and the 'Add Sources' window shown below opens. Select 'Add or create constraints' (2) and then select 'Next' (3).

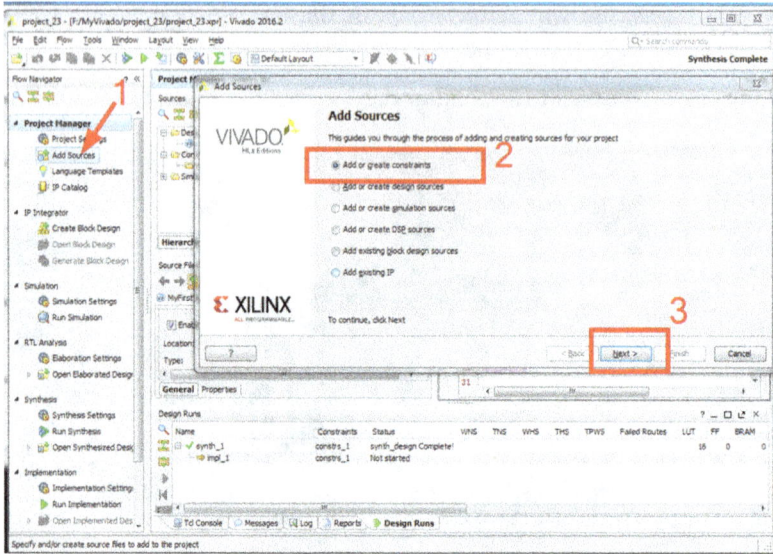

The window below opens; select either the plus sign or the 'Add Files' selection.

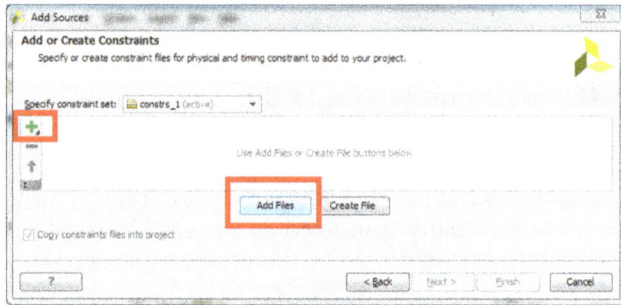

Select the 'BASYS3_Master.xdc' (1) file that you downloaded from https://github.com/Digilent/Basys3/tree/master/Resources/XDC and click 'OK' (2).

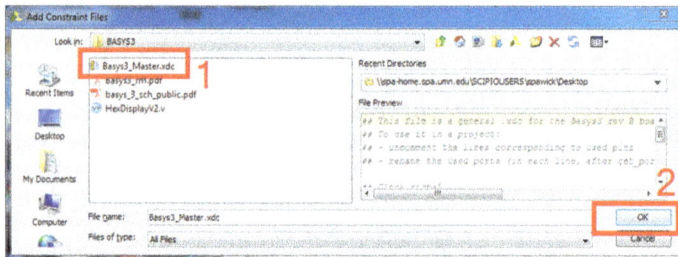

Before you close the next window shown below, confirm that a copy of the constraint file is added to your project (1) so that you can modify it. Click 'Finish' (2).

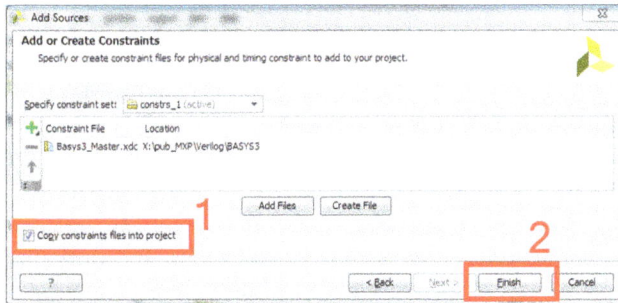

Expand the 'Constraints' folder in the 'Sources' panel until you see the newly added 'Basys3_Master.xdc' file (1). Double-click on it until you can see its contents in the window on the right, (2).

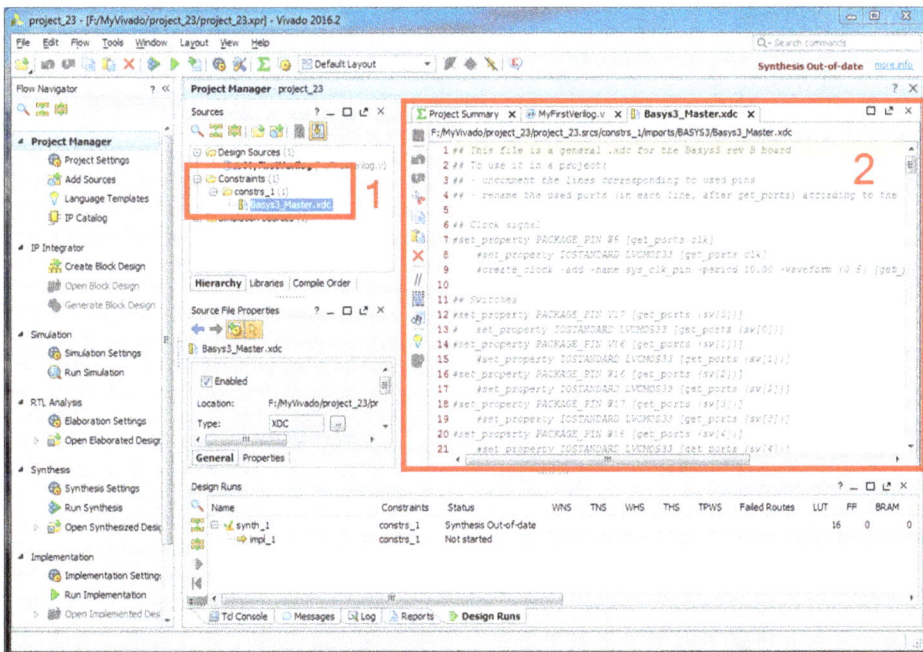

Take a moment and scroll through the file. First, you will notice easily identifiable sections pertaining to different hardware components on the BASYS3 board, such as the clock signal, switches, LEDs etc. Second, note that a specific component is always listed on two consecutive lines such as in the example shown below for led[0]:

```
#set_property PACKAGE_PIN U16 [get_ports {led[0]}]
#set_property IOSTANDARD LVCMOS33 [get_ports {led[0]}]
```

The first line corresponds to mapping the port led[0] to the correct physical pin, U16; the second sets the appropriate voltage level for the signal, LVCMOS33. Finally, note that the entire file content is greyed out, meaning that all the statements have been commented out. It will be your task to uncomment the required statements. However, before you do so, read the following.

There must be a one-to-one correspondence in Vivado projects between ports and constraints. In other words, there must be exactly one and only one constraint for each port. For obvious reasons, you will not be able to implement your design if you have ports specified in your Verilog code that have no physical connections to the 'outside' world because of missing constraints. For similar reasons, Vivado will (sometimes) fail to compile if you list physical connections in your constraint file that have no corresponding I/O ports in the Verilog code.

Keep in mind that the 'connection' between the physical ports in the constraint file and the Verilog's I/O ports is achieved by matching the names for the ports in the two documents. For example, it was no accident that you were instructed earlier to name your ports in the Verilog code 'sw' and 'led' because those are the names selected by the creator of the constraint file for these physical components. A list of some of the physical components with their constraint file names is shown in table 2.1. (*Note*: like everything else in Verilog, the names are case sensitive.)

(*An aside:* You are not restricted to these port labels in your Verilog code. Frankly, you may choose any descriptive name you fancy as long as you then *also* change its corresponding name in your constraint file so it is identical. However, for

Table 2.1. The table below lists the name (as used by the master constraint file), vector size, direction and brief description of the major components on the BASYS3 board.

Name	Bits	Input/output	Description
clk	1	Output	100 MHz system clock
sw	16	Output	Switches SW0 through SW15
led	16	Input	LEDs LD0 through LD15
seg	7	Input	Seven-segment display
an	4	Input	Anode for each seven-segment display
btnC	1	Output	Center button
btnU	1	Output	Upper-most button
btnL	1	Output	Left button
btnR	1	Output	Right button
btnD	1	Output	Down button
JA	8	Input/output	Connector on the side

simplicity, time saving and to avoid confusion, we will stick with the port names already provided in the existing constraint file.)

Based on the previous comments, you should now understand why the original constraint file had its entire content disabled. It is your task now to selectively enable the statements required to map your Verilog I/O ports, namely the ones pertaining to 'sw' and 'led'.

Start with the switches: highlight the entire section of the constraint file that concerns itself with the switches, i.e. starting with sw[0] and ending with sw[15], i.e. lines 12 through 43. (See (1) in the screenshot below.) We could now individually remove the hashtag symbols, #, at the start of each line to uncomment all these lines, although this would be very tedious. (*Note*: while Verilog uses the double backslash, //, to mark a comment statement, constraint files use a hashtag. Go figure...)

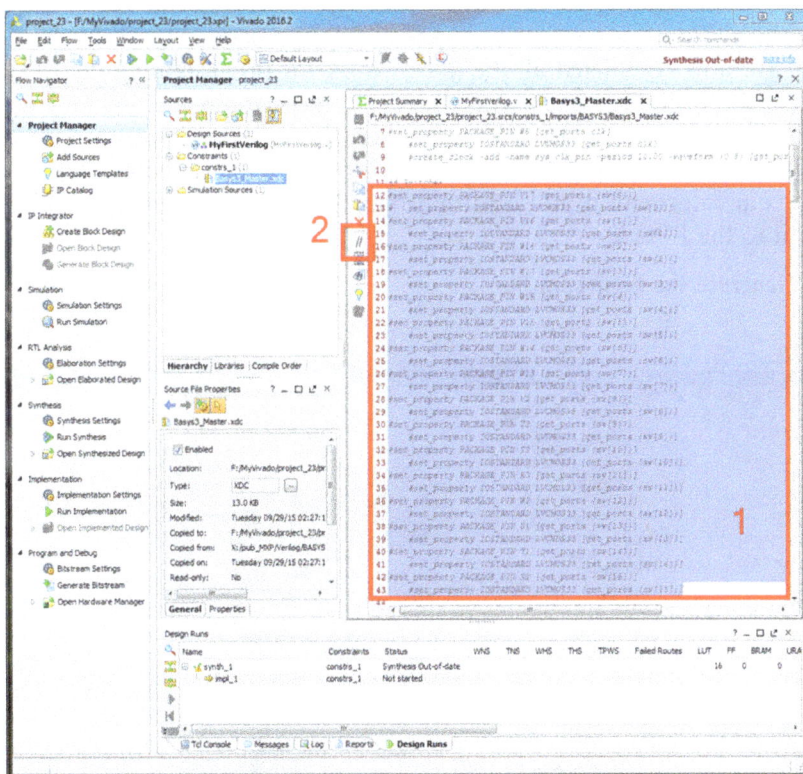

Instead, the same can be achieved much more conveniently by clicking on the 'Toggle Line Comments Tool' (2). Try it. Click on it and you should see all the greyed out text becoming 'active', i.e. color coded as shown in the screenshot below. (*Note*: if you keep clicking the tool, it toggles the highlighted text back and forth between greyed out and active.)

Now repeat this process for the LEDs. Select and highlight all the lines in the constraint file pertaining to the LEDs and use the toggle line comment tool to activate them.

You have now completed all the (hard) work for programming the FPGA and have created all the files where your input was needed. In the previous step, you specified the tasks for the FPGA in the Verilog source file and in this step you mapped the ports to the physical pins on the device. What remains is for Vivado to compile your information into a format that the FPGA understands and then to transfer it to the FPGA.

Step 4. Generate the bitstream file

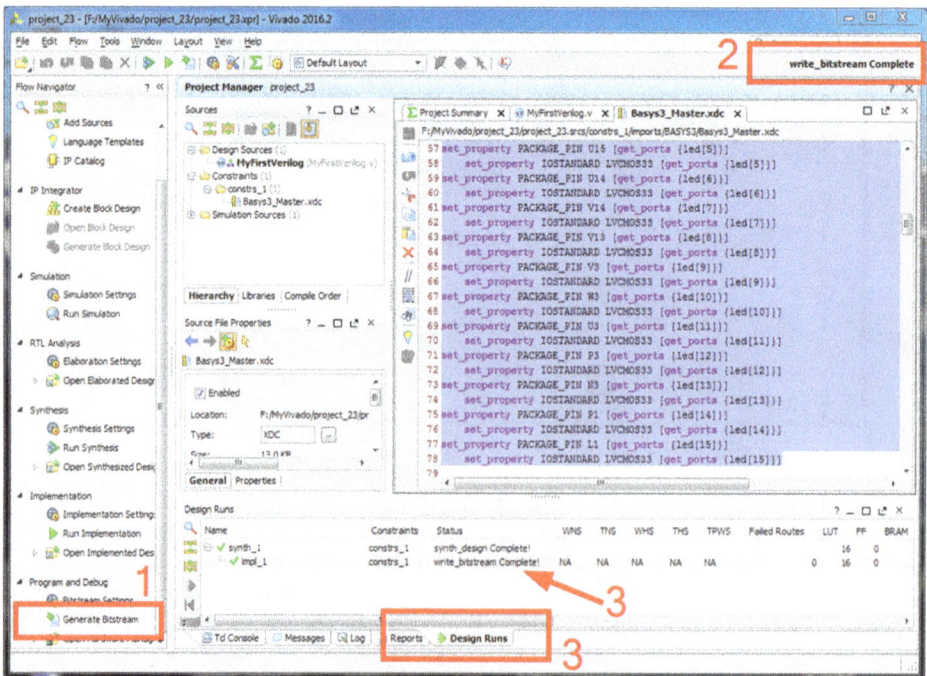

At this point, you are ready to generate the data file that will be used to program the FPGA. When Vivado compiles this file, known as a bitstream or firmware file, it will analyze your design, optimize it and then map each component to the actual hardware inside the FPGA. Depending on the complexity of the project, this may take a while, from a few seconds to a minute or so.

Start the process by selecting in the 'Flow Navigator' under 'Program and Debug/ Generate Bitstream' (see box 1 in the screenshot below). (If a window should pop-up asking you to save your information, do so; also, agree to the one informing you about having to run synthesis first.) To monitor the bitstream generation progress, pay attention to the status indicator in the upper right-hand corner (2) or select the 'Design Run' tab in the bottom panel (3).

A successful generation is marked by the 'write_bitstream Complete!' message in the status window. (See 2 or 3 in the screenshot above.)

Step 5. Programming the FPGA

In this last step, you can finally test your work. You will transfer the bitstream file to the FPGA and will have a chance to see if it works and whether or not you are able to control the LED with the switches on the board. Before you proceed, make sure the BASYS3 board has been connected to your computer and turned on.

Before you can program the FPGA, Vivado needs to find the target board and establish communication with it. Start by selecting 'Flow Navigator/Program and Debug/Open Hardware Manager/Open Target' (see 1 in the screenshot below). Select 'Auto Select' and a new window (3), opens up displaying the device to be programmed.

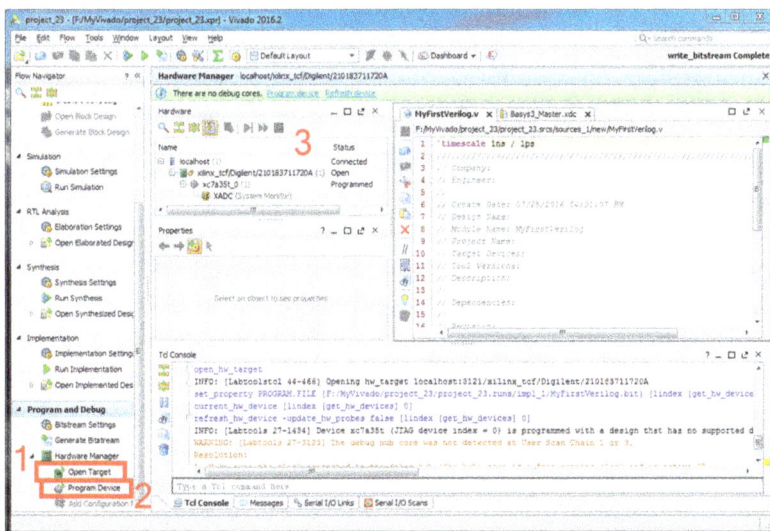

To program the FPGA, left-click on 'Program Device' (2) under 'Program and Debug/Hardware Manager' and select 'xc7a35t_0'.

Another window opens listing the name of the bitstream file which should be identical to the name you gave your module. Select 'Program'.

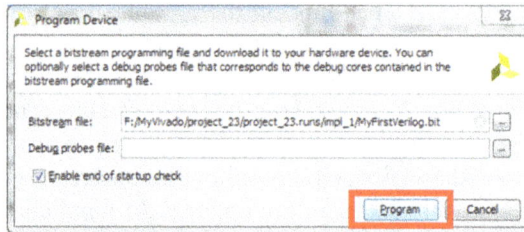

Congratulations, you are finally done. Test your design and operate the switches. Do the LEDs turn off when the switches are up and on when the switches are down? If so, congratulations! If not, check your steps.

Step 6. Summary

Now that you have successfully completed your first FPGA project let us review the steps and provide some additional comments.

First, in step 1, we created the project. Aside from providing a storage location for the numerous project files, this step also specifies the unique physical characteristics of the FPGA, such as its exact part number and package, i.e. number of pins.

In step 2, we listed the module's I/O ports and then had Vivado create the Verilog framework; we then added specific instructions by specifying the digital logic within the module, which in this case was a simple negate function. In a typical design, you will often return to this step to modify and fix the Verilog code and retest it.

In step 3, we mapped our Verilog I/O ports to pins on the FPGA and we also specified their voltage level by modifying an existing master constraint file. As an alternative, we could have edited the port names in the constraint file so that they match the ones specified in the Verilog file.

In step 4, we generated the bitstream file. You must regenerate this file each time you make changes to the Verilog code or the constraint file.

In step 5, Vivado acquired the FPGA target and programmed it. If you do not disconnect or power the board off, you can directly program it without having to re-acquire the target.

Finally, note that steps 2 and 3 could have been executed directly from step 1 when we were asked to provide or create the Verilog and constraint file. Since this is the first time using this software, we thought it better to go through each step separately and in great detail.

2.3.4 Review exercises

Since we have covered a lot of new material, all the exercises have been moved to the next section where we will cover additional Verilog concepts.

1. Create the Verilog source code from this section that negates the switch input and then displays the results with the LEDs.

2.4 Introduction to Verilog: wires, vectors, buses and Verilog operators

Before we can move on to more sophisticated digital designs, we will cover two fundamental Verilog topics, namely ports, wires and vectors, and Verilog operators.

2.4.1 Single-bit wires and vectors

When connecting (or assigning) components in a digital design we use either single-bit wires or vectors. As you recall from the last project, vectors are a collection of single-bit wires all having the same direction, i.e. they are either all inputs or outputs. They provide a convenient and efficient way to manipulate a large number of single-bit wires and they generally simplify the Verilog code. However, when working with wires and vectors we are often confronted to unbundle vectors to obtain access to a specific wire (or sub-collection of wires) or to bundle, i.e. to 'concatenate', single-bit wires into a vector. We will explore these two tasks in the following exercises after exploring the difference between I/O ports and internal wires.

2.4.2 I/O ports and internal wires

Every Verilog module contains I/O ports through which it communicates with external hardware such as switches, LEDs and buttons. They are always declared in the Verilog's module header. The I/O ports can consist of single-bit wires or vectors. See if you can identify them in the sample code shown below:

```verilog
module MySecondVerilog(
    input btnC,             //center button BTNC
    input btnD,             //bottom button BTND
    output [15:0] led
    );

    wire myxor;             //declare a single bit wire
    assign myxor = btnC ^ btnD;  //the ^ represents the Verilog XOR operator

    assign led = myxor;     //assigns single bit "myxor" to LSB of "led"

endmodule
```

This module uses a 16-bit vector 'led' and two single-bit input ports, 'btnC' and 'btnD', each connected to a single push button. (Alternatively, one could have employed the notation shown below for the single-bit ports although the abbreviated one shown above is typically used.)

```verilog
    input [0:0] btnC,       //center button BTNC
    input [0:0] btnD,       //bottom button BTND
```

Next notice the following declaration in the module shown above:

```verilog
wire myxor;
```

It represents an *internal wire*. It is different from an I/O port in that no external component can be directly connected to it unless the internal wire is 'somehow' connected to an I/O port. (Can you find in the example above where it is connected to an I/O port?) Typically, in simple designs such as the one shown above, there is not much use for such internal wires and it is mainly shown as illustration. However, as your designs become more complex you will have multiple components inside of your module and you will connect these with internal wires or vectors.

For a better example, study the module below. It acts identically to the first Verilog project in the previous section except that it uses an internal 16-bit vector called 'myvec' to create the negated switch values:

```
module MyThirdVerilog(

    output [15:0] led,
    input [15:0] sw
    );

    wire [15:0] myvec;          //declare a 16 bit vector
    assign myvec = ~sw;

    assign led = myvec;

endmodule
```

While declaring a single-bit internal wire is optional, declaring an internal vector is *mandatory*. (If you omit it, Verilog will default it down to a single-bit wire without a warning which is usually not what you intended.)

Now that you are familiar with the difference between I/O ports and internal wires, let us move on to extracting individual bits from vectors.

2.4.3 Extracting individual bits from vectors

To extract a specific bit of a vector x, use the following notation: $x[i]$, where i represents the ith bit. (When we talk about the ith bit we always assume that the least significant bit is the zeroth bit.) For example, in the code snippet below we permanently set the fourth bit of the (internal) vector 'myvec' high:

```
wire [15:0] myvec;
assign myvec[4] = 1'b0;   //1'b0 means 1 bit in binary (b) notation whose value is 0
```

(A quick remark on number representation in Verilog—we have more to say on that topic in the next chapter. The Verilog notation $x'by$ means that the value represented by y is in binary notation and that it takes up x bits. If the value of y takes up fewer bits than stated, Verilog will pad the result on the left with zeros; for example, 4'b1 will result in 0001. If the result takes up more bits, Verilog will truncate the higher bits; for example, 1'b1110 will result in 0.)

Or in the example below, we set *both* the third and fifth bit of 'led' to the value of the seventh bit of 'sw':

```
module MyFirstVerilog(
    input [15:0] sw,
    output [15:0] led
    );

assign led[3] = sw[7];
assign led[5] = sw[7];

endmodule
```

Instead of extracting a single bit, we can also extract a subset of bits from a vector by using the notation *x*[msb:lsb] where 'msb' represents the most significant bit and 'lsb' the least significant.

For example, below we assign the lower eight switches to the upper eight LEDs and vice versa.

```
module MyFirstVerilog(
    input [15:0] sw,
    output [15:0] led
    );

assign led[15:8] = sw[7:0];
assign led[7:0] = sw[15:8];

endmodule
```

Now that you know how to extract bits from a vector, let us look at the opposite problem, how to bundle individual bits into a vector.

2.4.4 Bundling bits into vectors, concatenation

Assume that we want to assign specific values to each component of a vector. For example, let us permanently turn every other LED on in a vector named 'led'. One way we could achieve this is using this code snippet:

```
assign led[0] = 1'b1;
assign led[1] = 1'b0;
assign led[2] = 1'b1;
assign led[3] = 1'b0;
```

This works, but shown below is a different notation using concatenation to obtain the same effect:

```
assign led = { 1'b0, 1'b1, 1'b0, 1'b1};
```

Concatenation works by assigning the right-most element in the list of values enclosed by the curly brackets to the least significant bit of the vector on the left-hand side; it then takes the next value to the left in the list and assigns it to the first bit, etc.

Of course, the values in the concatenation list could have been single-bit wires, vectors or components thereof, as in the messy example below:

```
assign led = {1'b1, 1'b0, 1'b1 , 1'b0, sw[0], sw[1], sw[2], sw[3], sw[15:8]};
```

Now that you are familiar with vectors and buses we leave a few examples for you to figure out at the end of this section. Some of them explore what happens when the bit size of the vectors do not match and how Verilog handles these.

2.4.5 Verilog operators

Table 2.2 shows the bit-wise and expression-wise Verilog operators. By convention, Verilog considers something true if any one bit in the expression is true. Of course, the inverse also holds: for something to be false, *all* bits must be low.

Table 2.2. Partial list of Verilog operators.

Bit-wise operators	Verilog notation
AND	&
OR	|
NOT (negate)	~
XOR	^
Right shift (by n bits)	$\gg$ n
Left shift (by n bits)	$\ll$ n
Expression operators	**Verilog notation**
Is equal	==
Is not equal	!=
AND	&&
OR	||
NOT (Negate)	!

Let us illustrate the difference between the bit-wise and expression-wise operators with the following two examples, whose only difference is in the operator in the assign statement:

```
//EXAMPLE 1

module MyBitWiseVerilog(
    input [15:0] sw,
    output [15:0] led
    );

assign led = ~sw;

endmodule
```

```
//EXAMPLE 2

module MyExpressionWiseVerilog(
    input [15:0] sw,
    output [15:0] led
    );

assign led = !sw;

endmodule
```

In example 1, in the 'MyBitWiseVerilog' module, the bit-wise negate operator, ~, reads each individual bit from the 'sw' vector, inverts it and assigns it to its corresponding bit in vector 'led'. We could have obtained the identical result in a more tedious manner by using the following Verilog notation:

```
assign led[0] = ~sw[0];
assign led[1] = ~sw[1];
assign led[2] = ~sw[2];
// and so on...
```

For example, if sw = { 1′b0,1′b1, 1′b0, 1′b1}, '~sw' would result in {1′b1, 1′b0, 1b1, 1′b0}.

In example 2, in the 'MyExpressionWiseVerilog' module, the negation acts on the entire expression, i.e. the 'sw' vector, as a whole and then operates on that result.

In other words the expression-wise operator first determines if the 'sw' vector is true or false by 'OR-ing', |, each individual bit:

$$X = sw[0]|sw[1]|sw[2]|sw[3]| \cdots |sw[15]$$

From this we can see that X can only be false when all bits of the vector 'sw' are low. In all other cases, X will be true. Next, since we applied an expression wide negation operator, the results of the previous operations will be inverted:

$$X = \overline{sw[0]|sw[1]sw[2]|sw[3]| \cdots |sw[15]}$$

This entire operation yields a single bit, which is then assigned to led[0]. For example, if 'sw = {1′b0, 1′b1, 1′b0, 1′b1}', '!sw' would result in 0.

Summarizing, a bit-wise operator acts on each vector component individually and will return a bit for each bit-wise operation acted on; an expression-wide operation acts on the entire expression or vector as a whole (by OR-ing each bit) and will return a single bit.

Enough theory, let us apply what we have learned.

2.4.6 Exercises

For the exercises you may either start each time with an entirely new project as outlined in the previous section, or simply reuse the project already created in the previous section and modify it. If you want to reuse the old project, open it and then select from the menu: 'File/Save Project As…'. This way you will create a copy of the previous project and you can always go back to it if you need to. Once you have modified the Verilog code, in the 'Flow Navigator' select 'Synthesis/Run Synthesis' and when that completes, run 'Program and Debug/Generate Bitstream' and then program the board in the 'Hardware Manager'.

For all these exercises, you should have to make changes only to the Verilog code, not to the constraint file generated in the previous section.

Exercise 1. Vector operations

In the first exercise, let us practice Verilog vector operations. Create a complete Verilog project that allows you to perform all these tasks. Specifically:

    a. Turn the four least significant LEDs (led[3], led[2], led[1] and led[0]) permanently on.

    b. Turn the next four LEDs (led[7], led[6], led[5] and led[4]) permanently off.

    c. Directly control the next four LEDs (led[11], led[10], led[9] and led[8]) through switch sw[0]. In other words, if sw[0] is on (off) these LEDs should be on (off).

    d. Invert the next four LEDs (led[15], led[14], led[13] and led[12]) through switch sw[1]. In other words, if sw[1] is on (off) these LEDs should be off (on).

(*Hint*: first implement the first task and generate the bitstream and program the board, and make sure that it works. Repeat this with the subsequent tasks.)

Exercise 2. Assigning vectors with different size

In this exercise explore what happens if you assign vectors with different number of bits to each other. Again, start with the simple project from the previous section, perform a 'File/Save Project As...' and modify the code so it matches the code shown below. Generate the bitstream and program the board. It might be easier if you start with all the bottom switches in the OFF position.

```
module MyFirstVerilog(
    input [15:0] sw,
    output [15:0] led
    );

    assign led = sw[7:0];

endmodule
```

    a. For the module shown above, explain what happens to the LEDs that were not explicitly assigned to a switch. Which are they and are these treated as having HI or LO inputs by default?

```
module MyFirstVerilog(
    input [15:0] sw,
    output [15:0] led
    );

    assign led[3:0] = sw[7:0];

endmodule
```

    b. For the module shown above, identify and explain what happens to the bits of the switch vector, 'sw', which exceed the size of the 'led' vector.

```
module MyFirstVerilog(
    input [15:0] sw,
    output [15:0] led
    );

    assign led[15:8] = sw[7:0];
    assign led[7:0] = sw[15:8];

endmodule
```

    c. Explain what the module shown above does.

Exercise 3. Bit-wise versus expression-wise Verilog operator

The two modules shown below demonstrate the difference between the bit-wise AND operator, &, and the expression wide AND operator, &&. Program the BASYS3 boards with each version and explain.

Module A:

```
module MyFirstVerilog(
    input [15:0] sw,
    output [15:0] led
    );

    assign led = sw&sw;        //bitwise AND
endmodule
```

Module B:

```
module MyFirstVerilog(
    input [15:0] sw,
    output [15:0] led
    );

    assign led = sw&&sw;  //expression wise-AND
endmodule
```

Exercise 4. Bit-wise Verilog operator left and right shift

The Verilog statement $x \gg n$ shifts a vector x to the right by n bits. Similarly, the statement $x \ll n$ shifts it to the left by n bits. Explore the shift operator and program the BASYS3 board with the module below.

```
module MyFirstVerilog(
    input [15:0] sw,
    output [15:0] led
    );

    assign led = sw<<4;
endmodule
```

Once programmed, turn all the switches (at the bottom of the board) off. Start with the least significant switch, SW0 and check which LED turns on. Next try the switches to the left of it, SW1, SW2 etc. Does it make sense? Explain what happens to the bits from switches sw[12:15]. Can you ever control led[3:0]?

Exercise 5. A simple XOR gate
In section 2.2, you used a discrete 7486 XOR chip which contained four XOR gates and you obtained its truth table. As stated earlier, this FPGA contains up to a million generic gates. In the Verilog module shown below, we implement one such XOR gate using the Verilog XOR operator, ^, and use switches sw[0] and sw[1] as inputs and led[0] as its output:

```
module MyFirstVerilog(
    input [15:0] sw,
    output [15:0] led
    );

    assign led[0] = sw[0]^sw[1];
endmodule
```

a. Test that it indeed works identically to the previously used 7486 XOR gate by comparing its truth table to the one obtained with the 7486 XOR gate.
b. An XOR operator is not a fundamental Boolean operator and it can be 'decomposed' into a combination of AND, OR and NOT operations. Modify the code shown above and use only the Boolean AND, OR and NOT operators to obtain the XOR functionality for led[1] while still using sw[0] and sw[1]. Replace the 'FILL IN YOUR OWN CODE HERE ???' statement shown below in bold with your implementation of the XOR operator using the AND, OR and NOT operators. Generate the bitstream and program the board. If you implemented it correctly, led[0] and led[1] should behave identically when operating switches sw[0] and sw[1].

```
module MyFirstVerilog(
    input [15:0] sw,
    output [15:0] led
    );

    assign led[0] = sw[0]^sw[1];

    assign led[1] = ??? FILL IN YOUR OWN CODE HERE ???;

endmodule
```

Exercise 6. Assignment statement
Below is a simple Verilog module. Its intent was to turn the LEDs on and off repeatedly, however, it will not compile. Can you explain why it will never ever, ever work? (*Hint*: it is not a syntax error; instead it contains a fundamental concept error involving the 'assign' statement and combinational logic. Remember that

Verilog describes a digital circuit so it may help to draw a schematic of the circuit below.)

```
module MyFirstVerilog(
    input [15:0] sw,
    output [15:0] led
    );

    assign led = sw;
    assign led = ~sw;
    assign led = sw;
    assign led = ~sw;
endmodule
```

2.4.7 Review Exercises

1. Create the Verilog source code that accomplishes all the tasks specified in exercise 1.
2. In exercise 2, identify in cases 'a' and 'b' the components of the vector which were not explicitly assigned and what their behavior is, i.e. do they remain HI or LO? Briefly explain what the module in part c does.
3. For exercise 3, the bit-wise versus expression-wise Boolean operators, explain for each of the two modules shown what happens to the LEDs when you operate the switches. Specifically, explain how many LEDs you can control with the first module and how many in the second module.
4. Briefly explain what the shift operator does in exercise 4. Describe what happens to the bits from switches sw[12:15]?
5. For exercise 5, show your Verilog code.
6. Explain why the code in exercise 6 will never work.

2.5 Instantiating modules

Instantiating a module means to implement an already existing module within your working module. It is similar to calling a function or subroutine: once you (or someone else) created the module you can use it repeatedly in your code by 'instantiating' it. Its benefits are to modularize your design, i.e. to build large projects by instantiating smaller and (hopefully) tried and tested components.

If that still sounds a bit abstract, think of the module that you will instantiate as a blueprint or set of instructions and specifications, for example the plans for a new building; 'instantiating' it creates a working 'copy' of the blueprint, i.e. the new building itself. Another analogy would be of a cookie cutter: once you or someone else has created the cookie cutter, you apply it repeatedly, i.e. you instantiate it each time, creating a delicious digital cookie.

In this section, you will create the module for a simple XOR gate and then instantiate it multiple times. Typically you would not create a module for a simple component such as a single XOR gate; however, we chose this exercise because of its simplicity. Once you are familiar with the process, you will use it to instantiate more complex modules.

2.5.1 Example of how to instantiate a module

In this exercise you will create a four-input XOR gate, as shown in figure 2.5.

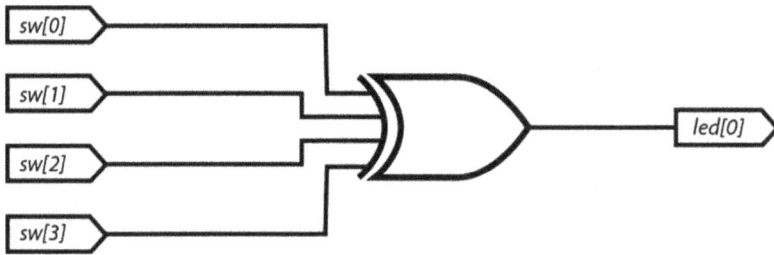

Figure 2.5. Schematic of a four-input XOR gate.

Such large multi-input XOR gates are employed in error correction algorithms to check that the transmitted number of HI bits is always odd or even.

Verilog operators are either unary, such as the negation operator, ~, or binary like the XOR operator ^, meaning they can 'operate' on only one or two operands at a time. So, creating a four-input XOR gate as shown above requires multiple applications of the operator on its product. One possible way to accomplish this is shown in figure 2.6.

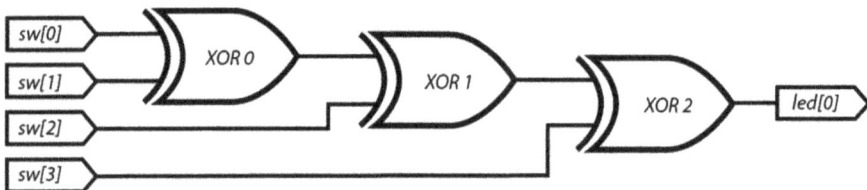

Figure 2.6. Implementation of a four-input XOR gate employing multiple two-input XOR gates.

We could implement this entire design in a single Verilog module. However, for this exercise we will create a separate XOR module and then instantiate it three times.

Start with an empty project similar to the exercise in section 2.3 and add the LED and switch inputs, so it looks like the one shown in the screenshot below (box 1).

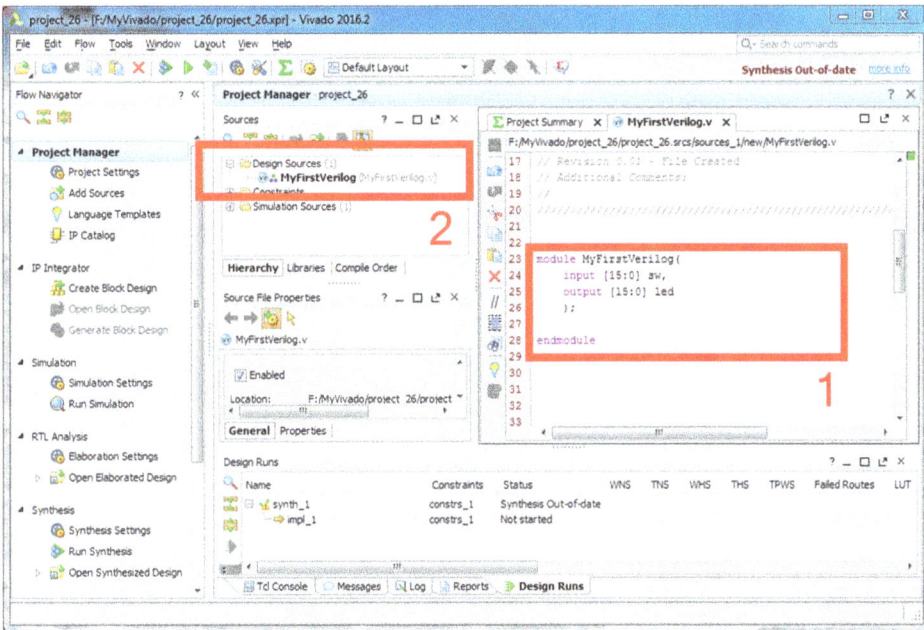

Before we are able to instantiate the three XOR gates within the module called (in the above example) 'MyFirstVerilog' we must first create a brand new and separate standalone module that will mimic an XOR gate like the one shown in figure 2.7.

Figure 2.7. Schematic representation of our two-inut XOR gate Verilog module. with inputs 'a' and 'b' and an output labeled 'q'.

You have two choices as to where you physically place the Verilog code for the new XOR module:

a. You can combine all the modules in a single file. For example, you could append the new code to the already existing file for the 'MyFirstVerilog' module. The order in which you place the individual modules within such a file does not matter.

b. Or you can keep each module in a separate file. In this case each file must be added to the project by selecting 'Project Manager/Add or Create Design Sources' in the 'Flow Navigator'.

Since the code for the new XOR module will be brief, let us use method 'a' and add another module below the existing 'MyFirstVerilog' module. The complete code for the new module, named 'MyXor', is shown below:

```
module MyXor(
    input a,
    input b,
    output q
    );
    assign q = a^b;
endmodule
```

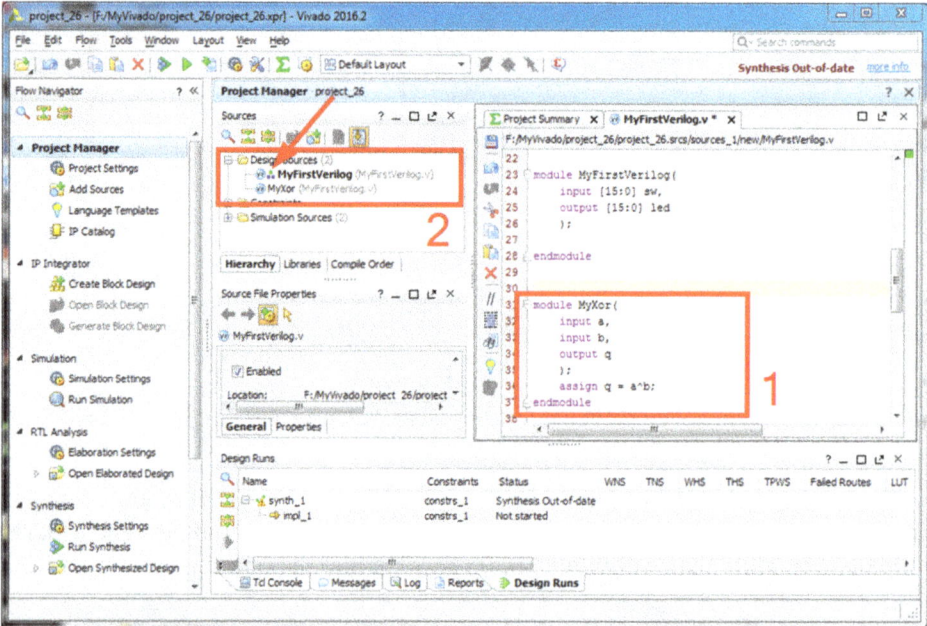

Before instantiating the newly created module, some comments:

- The 'MyXor' module uses single-bit I/O ports; there is no need for vectors.
- Currently there exist two modules within the project, 'MyFirstVerilog' and 'MyXor'. They are completely independent of each other and they are both listed in the 'Project Manager's Design Sources', see (2) in the screenshot above.
- With multiple modules in a project it is crucial that one (and only one) module is designated as the 'Top Module'. When you save your files, Vivado automatically picks (usually) the correct top module. This is indicated in the 'Project Manager/Sources/Design Sources' by preceding the module name by an icon of three squares; see the screenshot below where the 'MyFirstVerilog' module has (automatically) been chosen as the top module.

- You can override Vivado's choice for the top module: in the sources window, right-click on the name of your new top module and select 'Set as Top'. (This can come in handy when you are trying to debug the code and want to test a submodule by itself.)
- The purpose of specifying a top module is to associate the constraint file with the top module's I/O ports. (Submodules, i.e. instantiated modules never have constraint files associated with their I/O ports. Can you guess why? (*Hint*: think about multiple instantiations.) Also, the bitstream will only be generated for the 'Top Module'.)

Let us proceed and finally instantiate just one XOR module inside the 'MyFirstVerilog' module. The Verilog syntax for instantiation is:

```
OriginalModuleName InstantiatedModuleName(I/O port1, I/O port2, etc...);
```

In the screenshot below, we have instantiated the 'MyXor' module once (box 1). Note that the I/O port names during the instantiation (sw[0], sw[1] and led[0]) do not have to be identical to the ones used in the original module (a, b, q) but their order, number and direction must be preserved to match their purpose in the original module. This is equivalent to conventions in mathematical notation.

Once you have saved all the project files, 'File/Save All Files', observe how the source files were rearranged, (2), listing the instantiated module 'MyXorInst0', below the top module 'MyFirstVerilog'.

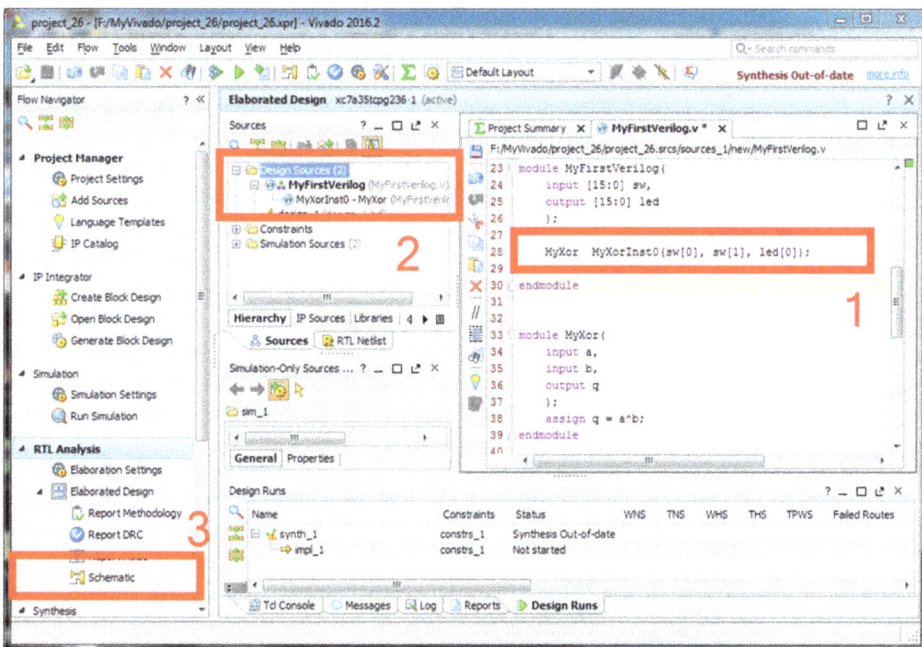

At this point we have really only created a single XOR gate. To illustrate this, or to check for any mistakes, let us try another useful feature of Vivado: we can display a schematic of your source code. Click on the 'Flow Navigator' on 'RTL Analysis/ Elaborated Design/Schematic', box 3 in the screenshot above. Next open the schematic window, see 1 below, and you will see the instantiated 'MyXorInst0' module inside of your top module.

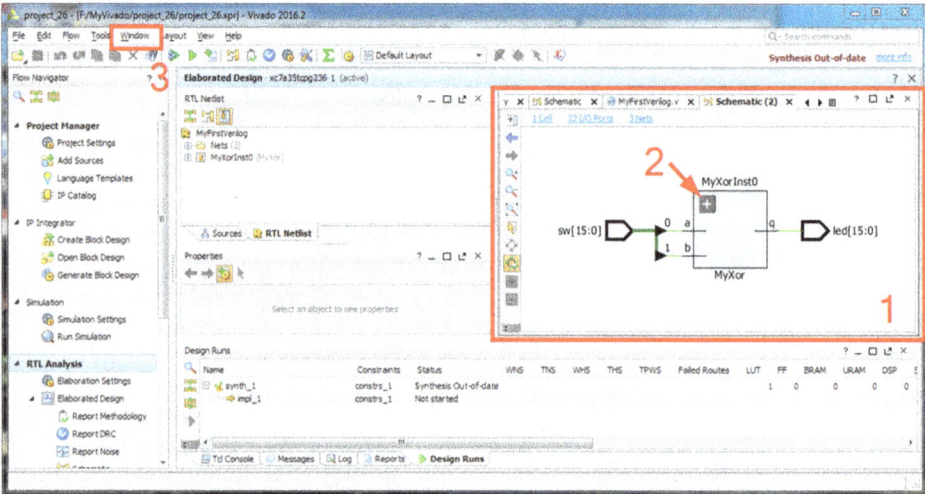

To display the content of the instantiated module click on the plus sign (2) and screen below opens revealing the (single) XOR gate (1).

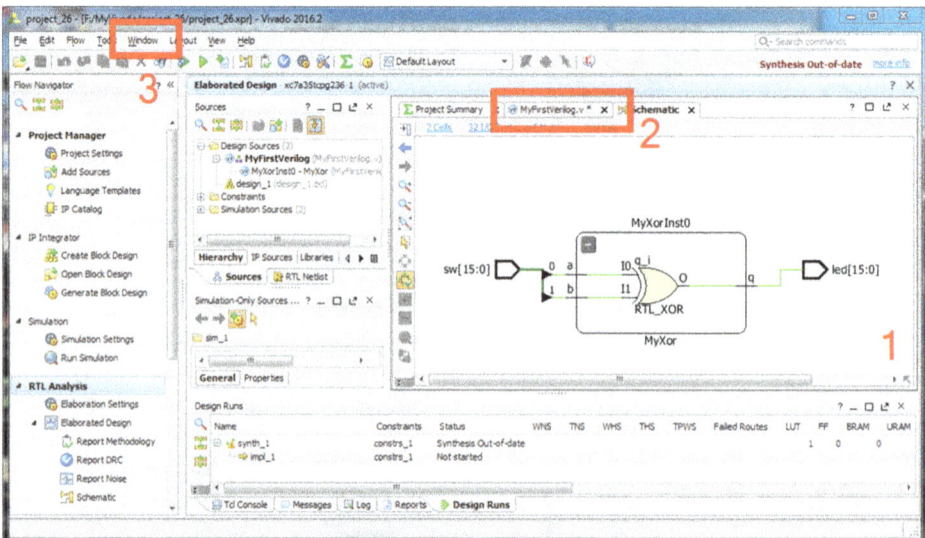

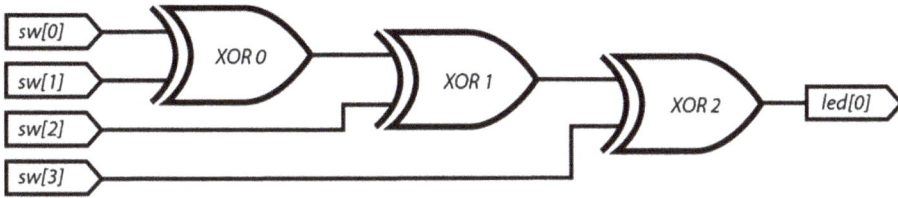

Figure 2.8. Schematic representation of our four-input Verilog XOR module. It was created by instantiating the previously created two-input XOR gate three times.

Let us go back to the source file and complete our four-input XOR gate design shown in figure 2.8 by instantiating two more XOR gates. Either click on the source file tab (2 above) or, on the source file under 'Design Sources'. (If you cannot see the source files then select from 'Menu Window/Sources'.)

The complete source code for the design is given below, with a screenshot and its schematic.

```verilog
module MyFirstVerilog(
    input [15:0] sw,
    output [15:0] led
    );

    wire temp0, temp1;

    MyXor   MyXorInst0(sw[0], sw[1], temp0);
    MyXor   MyXorInst1(sw[2], temp0, temp1);
    MyXor   MyXorInst2(sw[3], temp1, led[0]);

endmodule

module MyXor(
    input a,
    input b,
    output q
    );

    assign q = a^b;
endmodule
```

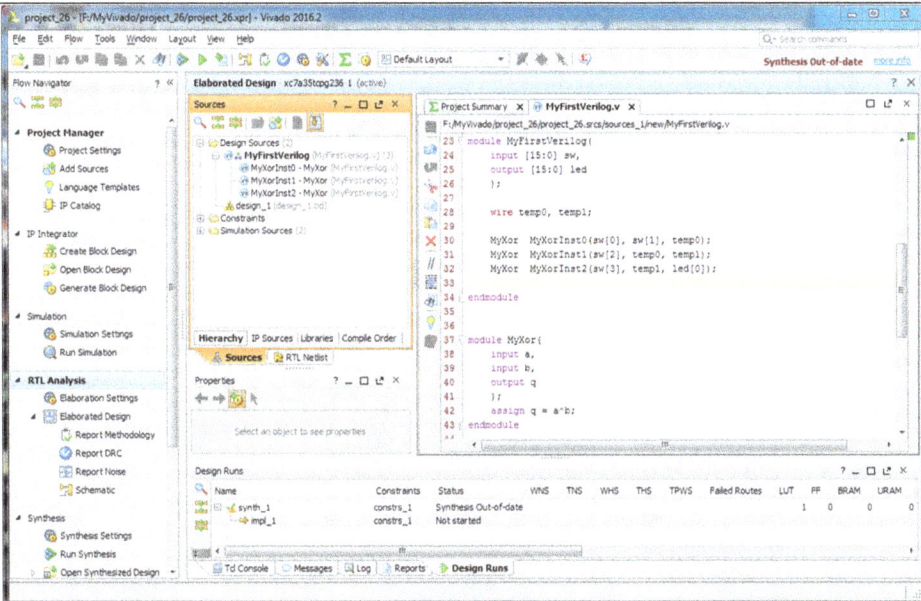

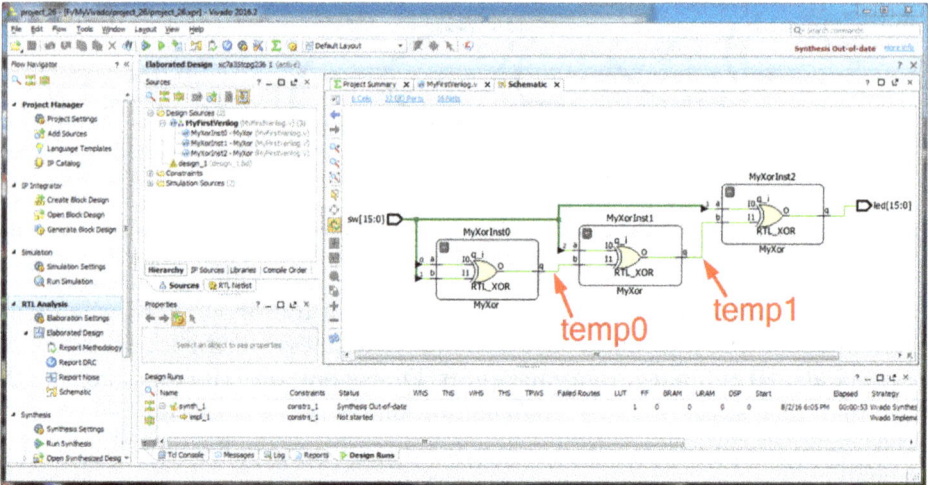

In the source code, as well as the schematic, note the explicit declaration of the internal wires 'temp0' and 'temp1'. They are required to connect the output of one XOR gate to the input of the next one. (*Note*: when you update your code and then try to view the schematic, you will get a message above the schematic window stating that the elaborated design is out-of-date. Select 'reload' and your schematic will be updated.)

2.5.2 Review exercises: eight-input XOR gate/parity checker

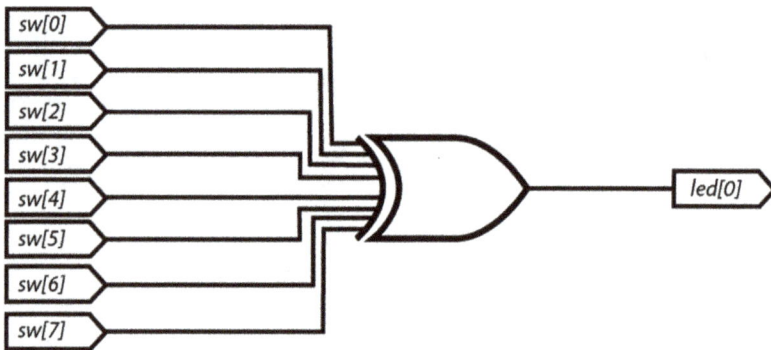

1. Expand your project to an eight-input XOR gate shown above. You may implement it in one of the following ways:
   - *Method 1.* Modify the code shown above and instantiate four more two-input XOR gates.

- *Method 2*. Modify the code shown above and create a four-input XOR gate module and then instantiate it twice together with one two-input XOR gate. The choice is yours.
  Create the Verilog code of your project as well as a print-out of the 'RTL Analysis/Elaborated Schematic'.

2. As previously mentioned, this type of circuit acts as an even or odd 'parity' checker to detect errors in data transmissions. It works by setting the parity bit, in our case led[0], to either HI or LO such that for any combination of 'sw' input bits the total number of HI bits (including the parity bit) will always be an even (or odd) number. This property is then used in (serial) data transmissions to check for corrupted data. First, the individual data bits, in our case the bits of the vector 'sw' are sent; second, the parity bit is appended to the transmission. (Typical transmission protocols also include some START and STOP bits to delimit the data package.) The receiver then counts the number of HI bits received in each package. If the number of HI bits disagrees with the previously agreed-on protocol, i.e. even or odd number of HI bits, the receiver then immediately knows when the data have been corrupted and it signals the sender to repeat the transmission.

3. Test your circuit built in 2.4.1. Use any combination of switch settings and count the HI bits, i.e. the switches that are turned on *including* the state of led [0]. Write three of these switch combinations down and conclude if this is an even or odd parity circuit.

4. Based on your answer to 3 above, can you convert the circuit into the opposite protocol, i.e. if it was an even parity circuit can you make it into an odd parity circuit, or vice versa? Explain how you would achieve that.

Chapter 3

FPGA and VERILOG: combinational logic part II

Additional reading

This chapter is (mostly) self-contained and continues where the previous one left off. Read the following pages from Horowitz P and Hill W 2015 *The Art of Electronics* 3rd edn (New York: Cambridge University Press) for the sections of this book indicated:
- p 838 for section 3.1.1.
- pp 707–8 for section 3.1.2.
- pp 724–7 for section 3.3.

3.1 Binary number representation and Boolean mathematics

For many people, digital logic is synonymous with computers and computing. With computer hardware and algorithms becoming ever more powerful, some scientists are beginning to wonder at what point in the future computers could become self-aware and how that would affect us. However, when we look into the opposite direction, the past, we can only marvel at how such a simple system based on two binary states manipulated by few Boolean logic operators could have become so ubiquitous.

Historically, the three components that made computers possible were:
1. The representation of numbers by binary representation.
2. Boolean operators and the insight that by using the appropriate Boolean operators, one could mimic simple mathematical operations.
3. Being able to implement the previous two concepts physically at a large scale and at high speed finally led to the development and success of computers.

In this and the following chapters we will look in detail at these three steps as we work towards building a very simple calculator, actually more of an adder, from scratch using only basic Boolean logic and our FPGA. Although this calculator is a far stretch from the computers you carry around with you every day, keep in mind that these sophisticated machines somewhere in their deepest 'guts' contain the circuitry that you will create in these exercises.

In this chapter we will use the numerical display on the BASYS3 board and look at binary number representation. In subsequent chapters, you will work on the Boolean logic that will mimic basic addition operations.

3.1.1 Exercise: Instantiating the seven-segment display

In this exercise you will instantiate an already existing Verilog module to utilize the four-digit HEX display on the BASYS3 board. You can download this Verilog module, named 'HexDisplayV2.v' from the site below: https://sites.google.com/a/umn.edu/mxp-fpga/home/vivado-notes/phys4051-course-related-materials.

You will control the digital values with the switches and display the resulting values in binary, decimal and hex using the LEDs and the HEX display. See the elaborated design schematic for the entire project in figure 3.1.

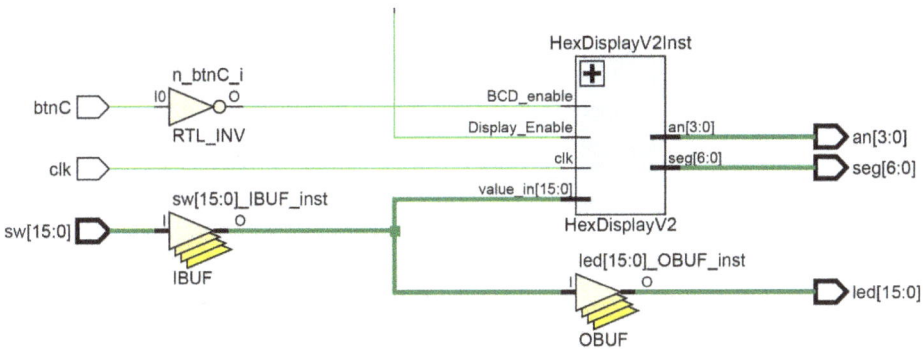

Figure 3.1. Elaborated design schematic of the instantiated HexDisplayV2 module to display the switch settings on the seven-segment display.

From your previous projects, you should recognize the buffered switch inputs, 'sw', and LEDs outputs, 'led'. (The buffers IBUF and OBUF are internal to the FPGA and protect its input and outputs.) The key component of the project is the instantiated HexDisplayV2 module and table 3.1 lists its I/O ports, their direction, size and purpose.

Table 3.1. HexDisplayV2 module port names with direction, vector size and description.

Name	I/O	Bus size (bits)	Purpose
clk	Input	1	Clock signal: 100 MHz BASYS3 board system clock; required for updating the display.
value_in	Input	16	Binary value to be displayed.
BCD_enable	Input	1	If this is set to HI, the display is set to decimal (or base 10) notation; otherwise it is hexadecimal, i.e. base 16 notation.
Display_Enable	Input	1	If it is high, the display is turned on; otherwise it is turned off. For this project set it high permanently.
seg	Output	7	Output signal to turn each of the seven segments of the display digits on or off.
An	Output	4	Output signal to turn one of the four displays on or off.

The other new element of the design is the 'btnC' I/O port and it corresponds to the center button of the five push buttons in the lower right-hand corner of the board. We will use its inverted output to control the display's numerical format.

The following are detailed directions on implementing the Hex display project.

1. Create a new project, create the source file, name the module and add all the I/O ports (with the correct name, size and direction) as shown in the diagram above.

2. Add the master constraint file and uncomment all lines pertaining to the I/O ports listed in the previous step. (For the clock signal, uncomment in addition to the first two lines also the line beginning with 'create_clock – add...'.)

3. Before instantiating the HexDisplay module, test your design by assigning the switches to the LEDs. Run 'Synthesis', 'Generate the Bitstream' and 'Program the Device' and check that you can indeed control the LEDs through the switches.

    If you have problems, consult the screenshot below and the schematic in figure 3.2.

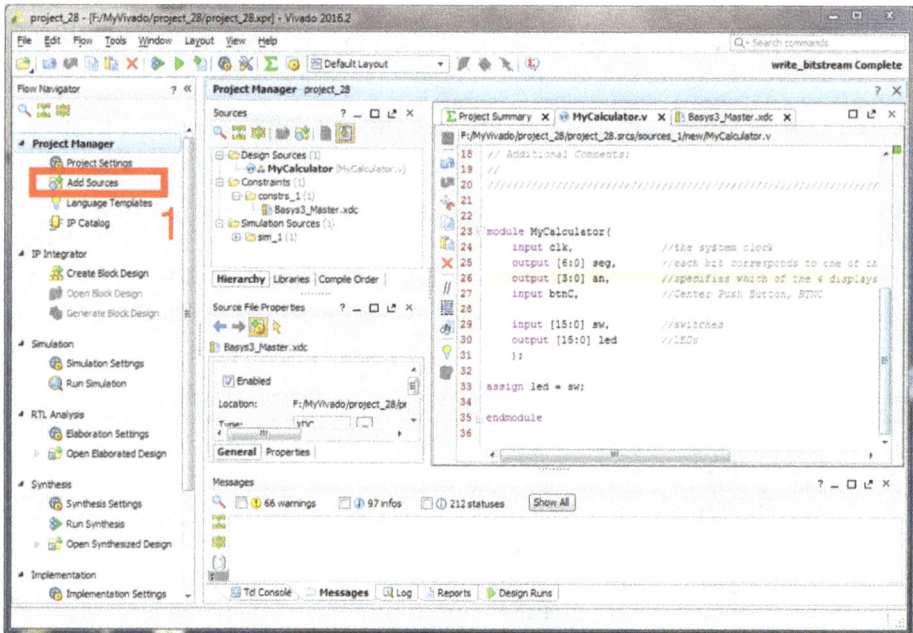

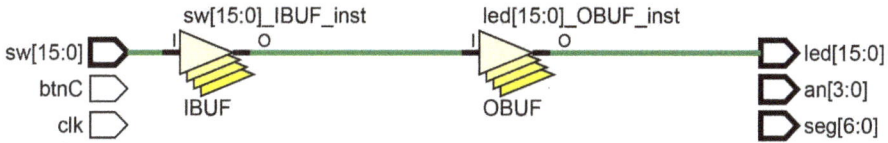

Figure 3.2. Elaborated design schematic of the top module prior to instantiating the HexDisplayV2 module.

Note the unused I/O ports in the schematic. We will make use of them in the next part when we instantiate the display module.

4. If you have not already downloaded the Verilog module named HexDisplayV2.v from the site below, do so now: https://sites.google.com/a/umn.edu/mxp-fpga/home/vivado-notes/phys4051-course-related-materials.

5. Before you can instantiate the display module, its source file must be added to your project. In the 'Flow Navigator/Project Manager/Add Sources' (1 in the previous screenshot) select 'Add or create design source/ + /Add Files…' and then select the HexDisplayV2.v file from the U:\pub_MXP\Verilog \BASYS3 folder. Confirm that 'Copy sources into project' (1) is enabled and hit 'Finish' (2).

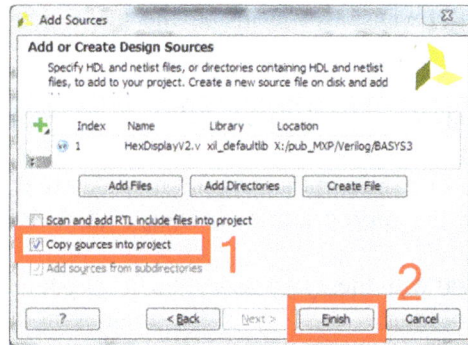

Back in Vivado's IDE, note the following. First, the new source file has been added to the 'Design Sources' (1). However, in doing so Verilog moved the 'Top Module' designation, indicated by the icon with the three squares, from the 'myCalculator' to the 'HexDisplay' module. (Since the two modules are still not linked it is really a toss-up which one ends up being the top module; in your version, Verilog may have left it alone.) We could force the 'Top Module' designation back by right-clicking on the 'MyCalculator' module and selecting 'Set as Top'; however, once we instantiate the 'HexDisplay' module inside the 'MyCalculator' module, Verilog will automatically take care of that. Also, note the submodules ('MyHex2BCD', 'MyEnableDigit' and 'MyDisplayDigit' modules) that the 'HexDisplay' module instantiates.

Second, when you click on the 'HexDisplayV2.v' tab (2) you can see the HexDisplay's code (3) whose information you will need when you instantiate it next.

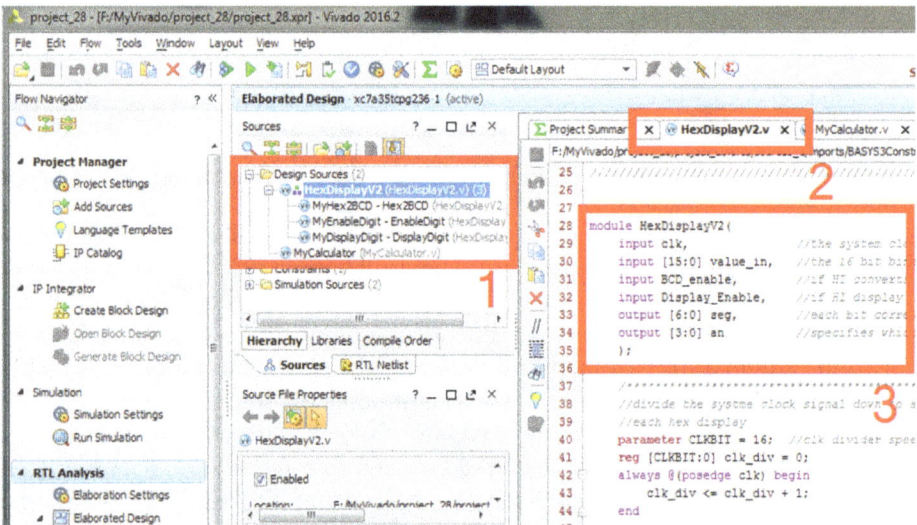

6. Instantiate the 'HexDisplayV2' inside the 'MyCalculator' module. When doing so, keep in mind the following two aspects:

    a. The order in which you list the individual ports or connections to the 'HexDisplayV2' module must adhere to the one used in the original module declaration.

    b. The I/O port names in the original module are dummy names; replace them with the appropriate ones from your design.

Let us see how we deal with these two issues. As far as point 'a' goes, you can get a sense of the order by looking at the code of the original module, shown as 3 in the screenshot above. As you can see, the first signal is a single-bit clock signal, then a 16-bit vector representing the value we want to display etc. All in all, the module has six I/O ports. You may recall that the syntax for instantiating a module (with three I/O ports) is:

```
OriginalModuleName InstantiatedModuleName(I/O port1, I/O port2, I/O port3);
```

An alternative way of writing this is to list each I/O port on a line by itself as in the example below:

```
OriginalModuleName InstantiatedModuleName(
   I/O port1,
   I/O port2,
   I/O port3
);
```

(One method to avoid confusing the order of the ports is to copy the entire module header of the original module (see 3 in the above screenshot) into the calling module and then selectively edit it. If you use this approach you must delete the keywords 'module' and all 'input' and 'output' designations. Also, you must add the new name for the instantiated module so it adheres to the syntax outlined above.)

As far as point 'b' above is concerned, you want to consult the complete elaborated design/schematic of our exercise, Figure 3.1, at the beginning of this section.

Working through HexDisplayV2 I/O ports in the order they are listed, we see from the schematic above that the top module's I/O port 'clk' is connected to the HexDisplayV2 'clk' port; in this case the names are identical and no change is needed. However, HexDisplayV2's next port, the 'value_in' vector is connected to the 'sw' vector; hence you need to adjust your instantiation statement to reflect this fact. So the first lines of the instantiation code look like:

```
HexDisplayV2 myHexDisplay(
        clk,
        sw,
```

The next HexDisplayV2 port, 'BCD_enable' is connected to the negated 'btnC' I/O port. Therefore, when you list the 'btnC' port in your instantiation you must precede it by the appropriate Verilog operator so it will be negated.

Since the 'Display_Enable' port is set high throughout, we will replace it directly with the constant 1'b1. Finally, the last names of the last two ports, 'an' and 'seg', are identical to the top module's port and do not need to be changed.

In case you run into trouble, here is the complete instantiated statement (with comments):

```
HexDisplayV2 myHexDisplay(
        clk,            //the system clock
        sw,             //the 16 bit binary value to be displayed
        ~btnC,          //if HI converts binary value into decimal value, else displays HEX
        1'b1,           //if HI display is enabled, LO turns it off.
        seg,            //each bit corresponds to one of the 7 segments on the display
        an              //specifies which of the 4 displays is to be turned on temporarily
        );
```

Once everything looks fine, select 'Synthesis/Run Synthesis' and the check in the 'RTL Analysis/Elaborated Design/Schematic' that it looks like the one in figure 3.1.

Generate the 'Bitstream' and program your board. Operate the switches and press the center push button, BTNC, and see that the HEX display and LEDs change.

3.1.2 Binary number representations and notation

Most of the numbers we will be working with are expressed either in decimal or binary notation. An (unsigned) binary number A, composed of n bits $A_{n-1} A_{n-2} \dots A_2 A_1 A_0$, is equivalent to the decimal value, z, through the following conversion:

$z = 2^{n-1} A_{n-1} + 2^{n-2} A_{n-2} + \cdots + 2^2 A_2 + 2^1 A_1 + 2^0 A_0$. Or, expressed in a more compact form

$$z|_{10} = \sum_{(i=0)}^{(n-1)} A[i]2^i.$$

Another numerical representation you will encounter is the base 16 or hexadecimal notation, or 'hex' for short. (In popular literature it is often written with a '0x' prefix such as 0x10, or an 'h' suffix as in 10h.) To express the 16 unique values that a hexadecimal number can take on, it was decided to augment the decimal numbers above 9 with letters. In other words, one counts in hexadecimals 0, 1, 2, 3, 4, 5, 6, 7, 8, 9, a, b, c, d, e, f, 10, 11…. (Typically, you will find either upper or lower case being used for the letters.) As you can see, 0xa represents decimal 10 and 0xf is decimal 15.

The conversion of a hex number B composed of $B_{n-1} B_{n-2} \ldots B_2 B_1 B_0$ to decimal is similar to the method used above but we have to take into consideration the base 16:

$$z|_{10} = \sum_{(i=0)}^{(n-1)} B[i]16^i.$$

The popularity of hex notation is due to two facts: first, it is exceedingly annoying to express large numbers in binary. For example, the (not very large) decimal number 1000 in binary is: 1111101000. Its equivalent representation in hex is much more compact: 0x3E8. Second, it is straightforward to convert between binary and hex and vice versa. For example to go from 0x3E8 to binary take each hex digit and convert it to its corresponding 4-bit binary digit, i.e. 3 is 0011, E is 1110 and 8 is 1000, i.e. 0x3E8 = 0011 1110 1000.

Using the correct base or radix notation is important in Verilog. If it is not specified, Verilog interprets any number as a decimal value. For example, the code snippet shown below would result in A being set to 0000 0011 1110 1000, the binary representation of the decimal value of 1000.

wire [15: 0]A = 1000;

To avoid ambiguities with radix notation, Verilog uses the following notation:
<size>'<radix>value
where:
- 'size' represents the number of binary bits the number is comprised of, not the number of hex or decimal digits. The default is 32 bits.
- ' a separator, single quote.
- 'radix' is the radix of the number, specifically:
  o 'b or 'B: binary
  o 'o or 'O: octal
  o 'h or 'H: hex
  o 'd or 'D: decimal
  o The default is decimal
- 'value' is the actual representation of the value in the previously specified radix. *Note*: an optional underscore character, _, can be used to improve legibility. For example, 0000_0011_1110_1000 is easier to read than 0000001111101000.

The examples below all assign the decimal value 13 in three different radix systems to some vector:

```
assign a = 4'b1101;    //binary
assign b = 4'hD;       //hexadecimal
assign c = 4'd13;      //decimal
```

3.1.3 Review exercises: binary number representation exercises

1. Operating the switches allows you to observe the numerical value of the switches in binary on the LEDs and in decimal and hexadecimal on the HEX display, depending whether or not you press the center push button, BTNC. Make a table with three columns and work your way from 0 to (decimal) 17 using the switches. In the first column enter the binary value, in the second the hexadecimal and in the third the decimal.
2. See if you can set the switches to display your favorite number which happens to be 4051. Of course, you could cheat and find its binary equivalent using your calculator or the web. However, for now just play with the switches until you get it. What is 4051's binary representation? How about its hexadecimal representation? (Remember, just push the center push button BTNC.)
3. As you will see in the chapters ahead, a very crucial characteristic of analog-to-digital interfaces is the number of bits, n, that they can resolve. It is directly related to the sensitivity or resolution of the device with the resolution as defined as $1:2^n$. For example, an 8-bit device has a resolution of $1:2^8$ or 1 in 256 parts. Work out how many bits you will need for resolutions of $1:100$, $1:1000$ and $1:10^6$.

3.1.4 Signed binary numbers: two's complement

Let us consider how one would represent signed binary numbers. Understanding this concept will be crucial in the next section when we convert our adding calculator into a subtracting one.

When devising a scheme to represent negative binary numbers, it must conform to the rules we are familiar with from algebra. For example, when adding a negative value with its positive value, the numbers must always add up to 0, i.e. $X + (-X) = 0$. It also should adhere to the fact that negating a number twice will return its original value, i.e. $-(-X) = X$. With binary numbers, these rules can be implemented by invoking a representation called two's complement. It works as follows:

1. Before you start, restrict all subsequent mathematical operations to n bits. In other words ignore the $n + 1$ bit that, for example, an addition operation might produce. (Do not worry if you do not understand all the details of binary addition; it will be covered in the next section.)
2. Take your n-bit number, X, and invert each bit, i.e. $\sim X$. This is called its *one's complement*.
3. Add 1 to $\sim X$. This is its *two's complement*. (Keep rule 1 in mind and discard the $n + 1$ bit this operation may produce.)

Here is an example using a 4-bit number, namely 3, i.e. 4'b0011. One's complement of 3 would be 4'b1100 and two's complement, or the binary equivalent of -3, is 4'b1101. Let us check if two's complement complies with the two algebra rules listed previously. First, adding $3 + (-3)$ in binary yields: $0011 + 1101 = 0000$. (*Note*: as mentioned, we always restrict our operations to n bits and have ignored the carry or fifth bit in this addition operation.) Also, applying two's complement twice to 4'b0011 returns it to its original value. Using the two's complement methods seems to satisfy the rules listed.

The range of a signed binary number includes positive and negative numbers. While an unsigned n-bit number's range spans from 0 to $2^n - 1$, a signed number's range is from -2^{n-1} to $2^{n-1} - 1$. For example, for $n = 4$, the range of unsigned numbers is from 0 to 15; the range of the signed numbers is from -8 to $+7$.

Implementing two's complement in Verilog is straightforward by using the negation operator $\sim$ and the addition operator, $+$. For example, if we wanted to convert the switch inputs, 'sw', to two's complement and assign it to some vector 'my2scomp' you would use the following notation:

```
assign my2scomp = (~sw) + 1;
```

3.1.5 Review exercises: binary number representation: signed numbers and two's complement

1. What is one's complement of the 4-bit number 4'b0000? What about 4'b0000 two's complement? Does that make sense?

2. You previously saw that an n-bit unsigned binary number A can be converted to its decimal equivalent using this equation:

$$z|_{10} = \sum_{i=0}^{n-1} A[i]2^i.$$

Does this equation work for signed binary numbers? Elaborate using the value for the 4-bit number -3 given above.

3.2 Boolean algebra and adder circuits

In the previous section we briefly mentioned that Verilog is capable of performing binary addition using the addition operator, $+$. Since the FPGA contains only basic logic gates, it implements this operation by applying a fundamental Boolean 'circuit' that will mimic this process. Such a circuit has broad applications and in this section you will design this circuit from scratch and then implement it and build a simple calculator capable of adding and subtracting integer numbers.

This addition circuit has been given its own name: the half or full adder circuit. Unfortunately, this name choice can be misleading since the circuits have nothing to do with the mathematical concept of 'half' and it is a bit more obscure what 'full' refers to. A far better choice would have been to name the half adder a 2-bit adder and the full adder a 3-bit adder. However, the conventional names stuck and are what everyone else uses, and so will we.

3.2.1 Half adder circuit

Before we work out the Boolean logic of the circuits, let us perform a quick review of binary math. Luckily it adheres to the same rules that have been drilled into you since elementary school: $0 + 0 = 0$, $1 + 0 = 1$, $0 + 1 = 1$. In binary mathematics, the maximum value which can be expressed with a single bit can never exceed 1. Hence it follows that the operation of $1 + 1$ will require a 2-bit result, i.e. 10, the binary equivalent of 2.

From this we can conclude that adding two (binary single-bit) terms, $a + b$, will always produce an answer that can be expressed by a 2-bit vector, $q[1:0]$, i.e. $a + b = q$. Typically, the least significant bit, $q[0]$, is called the 'sum' bit, and the most significant, $q[1]$, the 'carry' bit. In summary, this is what the half adder (or 2-bit adder) circuit does: it adds two bits and produces a 2-bit answer.

You will implement this circuit by creating two truth tables. In the first, determine the values for the 'sum' bit output for all combinations of the (single-bit) inputs a and b; in the second, determine the values for the 'carry' output for all combinations of the (single-bit) inputs a and b.

Next, express the tables with Boolean logic expressions. In other words, find sum (a, b) and carry(a, b).

Implement your half adder by creating an independent Verilog module similar to the one shown in the schematic in figure 3.3.

Finally, instantiate your half adder in the top module created in the previous section. Use the two right-most switches of the BASYD3 board as inputs and display the output of the half adder in the hex display. Specifically, assign sw[0] to the input a of your half adder module and sw[1] to b. Create a 16-bit internal (vector) 'wire' [15:0] q and assign it to the 'value_in' input of the instantiated 'HexDisplay' module as shown in the schematic in figure 3.4. Assign the outputs of the half adder to the vector q; specifically, assign 'sum' to $q[0]$ and 'carry' to $q[1]$.

Test your circuit and operate switches SW0 and SW1 and check that the Hex display shows the correct addition result.

3.2.2 Full adder circuit

As you already have learned in elementary school, there exists a very simple algorithm that breaks down even the most complex addition problem into a repetition of simple steps, each involving the addition of three numbers at a time. The procedure that you were taught for decimal numbers applies equally well to

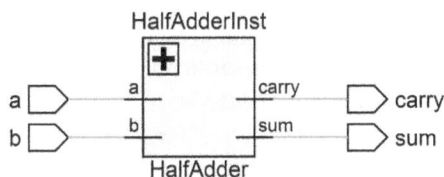

Figure 3.3. Schematic representation of a half adder module which is capable of adding two bits at a time.

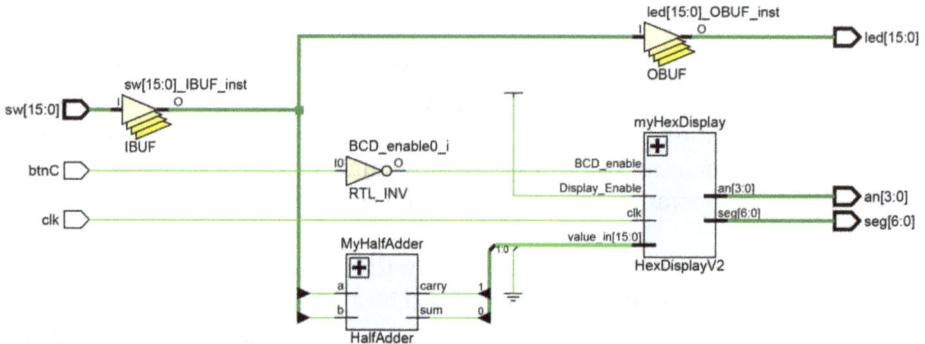

Figure 3.4. Elaborated schematic of the complete half adder project with the instantiated HalfAdder and the HexDisplayV2 module. Note that only switches SW0 and SW1 are connected to the 'a' and 'b' inputs of the HalfAdder.

Table 3.2. Detailed illustration of the binary addition process of the two decimal numbers 9 and 15.

	Decimal values	Binary values			
		Bit 3 (MSB)	Bit 2	Bit 1	Bit 0 (LSB)
$A[3:0]$	9	1	0	0	1
$B[3:0]$	+15	1	1	1	1
Carry bits $C[4:0]$		1*	1	1	1
	–	–	–	–	–
$Q[4:0]$	24	1**	1	0	0	0

binary numbers. (If you prefer, you could substitute the word 'bits' for 'digits' in the example shown below and it still would work.)

To illustrate the binary addition algorithm, consider the example shown in table 3.2. It utilizes the addition of two terms represented by vectors A and B, each four bits wide, with (arbitrary) assigned values of 9 (binary 1001) and 15 (binary 1111).

The first step of the procedure begins with the addition of the least significant bits, $A[0] + B[0]$. In the example shown above this is the addition of $1 + 1$, which results in a sum bit $Q[0]$ of 0, and a carry bit of 1. As you will see in the next section, it is advantageous to assign this carry bit to the vector component, $C[1]$ (instead of $C[0]$). (Note that this step could have been carried out with our previously designed half adder.)

The next step, and all subsequent ones, involves the addition of the next-higher bits, i.e. of the two bits to the immediate left *and* the addition of the carry bit from the previous step. In the example above, this corresponds to the addition of $A[1] + B[1] + C[1]$, i.e. $0 + 1 + 1$, resulting in a sum bit $Q[1]$ of 0 and a carry bit of 1 which we will again assign to the (next higher) carry bit, $C[2]$. The addition continues until all bits have been added and ends with the final operation of assigning $C[4]$ to the MSB sum bit, i.e. $Q[4]$.

It should be obvious how one can apply this addition algorithm to arbitrary large vectors and numbers: by moving from the least significant to the most significant bit we determine at each bit position i the sum bit of the 3-bit addition $A[i] + B[i] + C[i]$ and assign it to $Q[i]$; we also determine the corresponding carry bit and assign it to $C[i+1]$. The key component of this operation is the 3-bit adder which is the circuit known as the full adder. (*An aside:* you can always turn a full adder into a half adder by setting one of its inputs permanently to 0.)

Let us start implementing the Boolean logic for the full adder by producing complete truth tables for it. Similar to the half adder, find again the sum and carry bits as a function of inputs a, b, c as shown below, i.e. find Boolean expressions for sum(a, b, c) and carry(a, b, c).

Implement your full adder by creating an independent Verilog module similar to the one shown in the schematic in figure 3.5.

Finally, instantiate your full adder in the top module created in the previous section. (Remove or comment the half adder out.) Use the three right-most switches as inputs and then display the output of the full adder in the hex display. Specifically, assign sw[0] to input a of your full adder module, sw[1] to b and sw[2] to c. Create a 16-bit internal (vector) wire [15:0] q and assign it to the 'value_in' input of the instantiated 'HexDisplay' module as shown in the schematic in figure 3.6. Assign the outputs of the instantiated full adder to the vector q; specifically, assign sum to $q[0]$ and carry to $q[1]$.

Test your circuit and operate switches SW0, SW1 and SW2 and check that the Hex display shows the correct addition result from any combination of the three

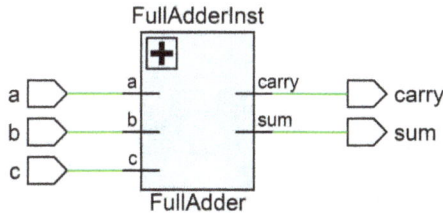

Figure 3.5. Schematic representation of a full adder module capable of adding three bits at a time.

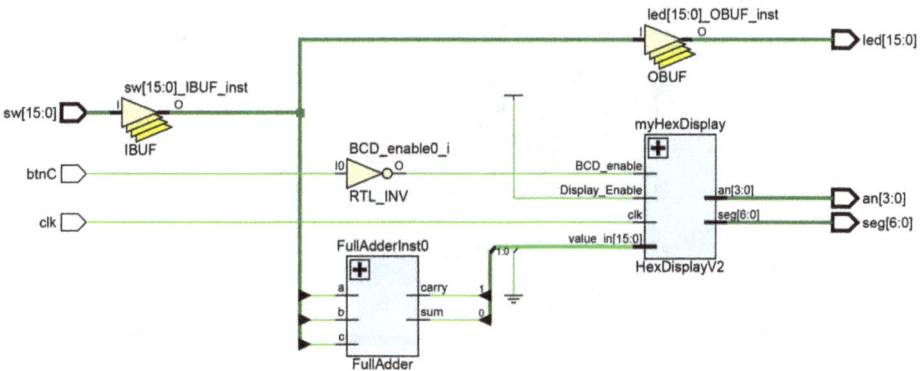

Figure 3.6. Elaborated schematic of the complete full adder project with the instantiated HalfAdder and the HexDisplayV2 module. The FullAdder's inputs 'a', 'b' and 'c' are connected to switches SW0, SW1 and SW2.

switch settings. Create your Verilog code and schematic with your instantiated full adder expanded. *Important*: save an additional copy of this program for later, i.e. use 'Save Project As…'.

3.2.3 8-bit adder/calculator

Let us now put the 16 switches on the BASYS3 board to good use and build a simple calculator capable of adding two 8-bit numbers A and B by instantiating the previously created full adder eight times (similar to the diagram in figure 3.8).

You should be able to complete the project on your own. However, if you require more detailed instructions (including the complete schematic) see below.

1. Use the code from the previous full adder project and declare two internal 8-bit wire vectors named A and B; assign them to the switches as shown in figure 3.7. Specifically, assign sw[7:0] to vector B and sw[15:8] to A. (This step is optional but it will greatly clarify subsequent assignments for the full adder inputs.)

2. Declare two internal 16-bit wire vectors named C and Q. You will use them for the carry and sum bits. Assign the sum bits, vector Q, to the 'value_in' input of the 'HexDisplay' module.

3. Now start with the calculator's least significant bit. Modify your previous code and set the inputs to your full adder to $A[0]$ and $B[0]$. Since you will need a half adder for the least significant addition bit, set the full adder's carry input, c, to 0. As far as the outputs are concerned, assign the sum bit to $Q[0]$ and the carry to the next higher carry bit, i.e. $C[1]$. You may want to test

Figure 3.7. Picture of the BASYS3 board illustrating how switches SW0 through to SW7 will represent vector B, while switches SW8 through to SW15 correspond to A.

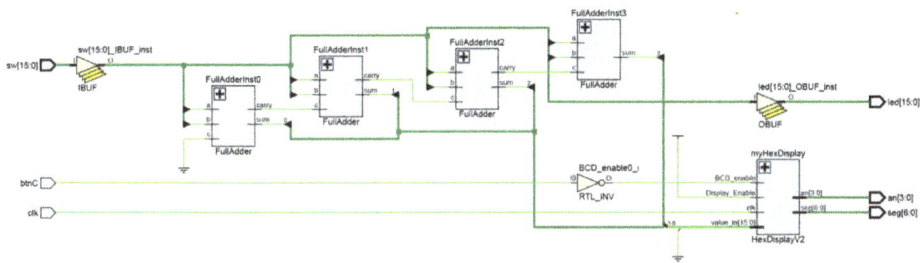

Figure 3.8. Complete elaborated schematic of a simple calculator. For the sake of clarity, the implementation shown is only capable of adding two four-bit vectors. (Your final design will be able to add two eight-bit vectors.)

your code to make sure it still works and the Hex display correctly shows the addition results of SW0 and SW8. (*Warning*: you will get the wrong answer for 1 + 1. Why?)

4. Add the next higher bit: instantiate another full adder and set its input to A [1], B[1] and C[1]. Similarly, assign its outputs to Q[1] and C[2]. You may want to test your code again before proceeding.

5. Continue in the same fashion and keep instantiating more full adders. Assign their inputs to $A[i]$, $B[i]$ and $C[i]$ and the outputs to $Q[i]$ and $C[i + 1]$ where i represents the particular bit you are working with.

6. Continue all this way up to the eighth bit, i.e. A[7], B[7] and C[7]. Again assign the sum to Q[7] but what should you do with the eighth carry bit? If you cannot figure it out, check the complete diagram in figure 3.8.

3.2.4 8-bit subtractor/calculator and two's complement

Let us now turn our calculator into one capable of subtracting two numbers, for example $A - B$, a feat (as you will see at the end of this section) that has far more profound applications than the simple mathematical operation that it performs. At this point we could proceed in two different ways. One method would be to work out the details of the full subtractor circuit and substitute it for our full adders. However, there is an alternative method invoking the fundamental algebra concept that any subtraction can be decomposed into the addition of a negative number. For example, $A - B$ is identical to $A + (-B)$. Since you learned in section 3.1.4 how to obtain the negative value of a binary number by invoking two's complement, we will use this approach. This will allow us to keep reusing most of the full adder project from the previous exercise.

Let us now modify your previous 8-bit adder project so that it is capable of subtracting $A - B$. You will need only two modifications to your Verilog code:

1. Modify vector B's switch assignment statement so that it reads as two's complement:
   ```
   assign b = (~sw[7:0])+1;
   ```

2. Limit the output of the adder operation (for the hex display) to the eight sum bits, q[7:0] i.e. ignore the MSB carry bit, C[8].

Synthesize your new project and program your BASYS3 board. Check that it subtracts and correctly displays $A - B$. Make life easy for yourself and try $8 - 2$.

Important: as mentioned in the previous section, for n bits, the range of numbers which can be represented by two's complement spans from -2^{n-1} to $2^{n-1}-1$. If the operation of $A + (-B)$ exceeds this range, then the algorithm no longer works. For simplicity, we will constrain all our exercises and review exercise questions to cases where the results will be within the stated range for two's complement.

3.2.5 Summary

As mentioned previously, Verilog is capable of performing (integer) addition, subtraction and even multiplication. However, what we wanted you to see is that underneath the mathematical operators exists a rather simple circuit. It relies on a few basic Boolean logic rules which you were able to derive and you could have built the circuit using just a few programmable switches, such as transistors.

The subtractor circuit with its two's complement operation is in itself pretty ordinary. However, its implications are actually far more subtle: this circuit allows us to compare two binary (integer) numbers and determine their relationship to each other. By examining the result of the subtraction we can determine which of the two numbers is larger or if they are identical.

A programmable circuit capable of adding, subtracting and comparing (and ideally multiplying and dividing) two numbers is called an arithmetic logic unit (ALU). It is the core of any central processing unit (CPU) found in a computer. It is often assumed that the operations or instructions a CPU can perform are very sophisticated. In reality, especially in reduced instruction set computers (RISC), they are very similar to the ones you have implemented in this project. However, by executing these instructions at extremely fast repetition rates and with the aid of clever algorithms, circuits very similar to the one you built control everything, from entertaining video games to vast infrastructures.

However clever our design was, it lacked two key components found in every computer. First, it was unable to store data. We will fix that in the next chapter by using flip-flops as memory. Second, it was not able to selectively switch between circuits. For example, it would have been nice if we could have selected, for example with a push button, if we wanted to add or to subtract our two numbers. You will learn how to implement such a selector circuit in the next and final section on combinational logic using multiplexers and demultiplexers.

3.2.6 Review exercises: 8-bit adder/calculator

1. Work out the truth table and Boolean expression for your half adder circuit. Create the expanded schematic and Verilog code for the complete half adder circuit project.

Table 3.3. Complete the table below and work out the decimal values the binary bits on the left represent, depending on whether we interpret them as signed or unsigned values.

Binary value	Unsigned decimal representation	Signed decimal representation (using two's complement)
000		
001		
010		
011		
100		
101		
110		
111		

2. Work out the truth table and Boolean expression for your full adder circuit. Create the expanded schematic and Verilog code for the complete full adder circuit project.
3. Create the schematic and Verilog code for the 8-bit adder/calculator project. What is the range of the calculator?
4. Complete table 3.3 and list the corresponding *decimal* representation for these 3-bit numbers. (*Hint*: the signed numbers have a range from −4 to +3.)
5. Create the schematic and Verilog code for the 8-bit subtractor/calculator project. What happens to the hex display if $A < B$? (Assume that all results of the $A + (-B)$ operation are within the two's complement range.)
6. Explain how can you detect if $A \geqslant B$? (*Hint*: which bit would you monitor?) Assume that all results of the $A + (-B)$ operation are within the two's complement range.
7. How would you detect if $A < B$? Assume that all results of the $A + (-B)$ operation are within the two's complement range. (*Hint*: use 6 above.)

3.3 Multiplexers (MUX) and demultiplexers (DEMUX)

For the last combinational logic circuit we will work with the versatile MUX/DEMUX circuit. It is a programmable switch circuit and is used in encoders, decoders and look-up tables (LUTs). After you have worked through this section you will appreciate the circuit's power in implementing designs, from simple gates to complete CPUs.

A n-to-1 MUX contains n signal inputs which are connected through programmable switches to a single output. At any given time, only one of the signal inputs is enabled and any changes this input receives will be directly forwarded to the single output. Similarly, changes to the disabled inputs will have no effect on the output. The enabled input is specified by the state of the selector lines, labeled as 'sel' in figure 3.9. Their decimal representation corresponds directly to the index of the enabled input. For example in the case of 4-to-1 MUX, setting the selector lines 'sel' to 2'b11, i.e. decimal 3, specifies that input 3, $a[3]$, is 'switched through' to output q;

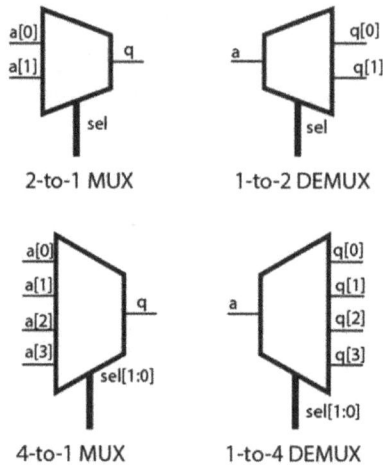

Figure 3.9. Schematic symbols of multiplexers, MUX, (on the left) and demultiplexers, DEMUX, (on the right).

all the other inputs will be ignored. As an analogy, a MUX acts like a soda machine with different types of soda at its inputs a[1]; you select the type of soda through the selector lines and it then appears at its output q.

The DEMUX circuit is the 'reverse' of a MUX. It has only one input but *n*-outputs. Again, its selector lines specify which output will carry the input signal while all the inactive or disabled output lines will be set by default to either high, or more typically, low.

A simple analogy to a DEMUX circuit is a phone system. By dialing a number, i.e. setting the selector lines to a specific value, the caller, i.e. the single input, is connected to one specific output among the many available ones. However, this analogy fails when we consider the directionality of the signal because it allows for the transmission of two signals simultaneously in opposite directions, i.e. it acts in a duplex mode. In a DEMUX (or MUX) the signal can only propagate from the input to the output and never the other way around. In other words, a DEMUX (or MUX) always acts in a simplex (transmission) mode.

See the diagrams in figure 3.9 and the corresponding simplified truth tables of table 3.4 for the MUX and DEMUX.

3.3.1 MUX application: look-up tables (LUTs)

To illustrate the versatility of a MUX, let us go back to the full adder example from the last section. If you have poked around Vivado, you may have noticed that you can display the schematic of your Verilog project in two different ways. First, there is the 'Elaborated schematic', showing you the Boolean logic implementation of your design.

However, there exists a second schematic of your project under 'Synthesized Design'. This synthesized schematic is a representation of your (highly) optimized design and it may have very little resemblance to your elaborated schematic. Of course, it will still work the same.

Table 3.4. Truth tables for the multiplexers (left) and demultiplexers (right) shown in figure 3.9. Note that we assumed the unused outputs of the DEMUX to remain low.

2-to-1 MUX		1-to-2 DEMUX	
sel	q	sel	q
0	$a[0]$	0	$q[0] = a$
			$q[1] = 0$
1	$a[1]$	1	$q[0] = 0$
			$q[1] = a$

4-to-1 MUX		1-to-4 DEMUX	
sel	q	sel	q
0	$a[0]$	0	$q[0] = a$
			$q[1] = 0$
			$q[2] = 0$
			$q[3] = 0$
1	$a[1]$	1	$q[0] = 0$
			$q[1] = a$
			$q[2] = 0$
			$q[3] = 0$
2	$a[2]$	2	$q[0] = 0$
			$q[1] = 0$
			$q[2] = a$
			$q[3] = 0$
3	$a[3]$	3	$q[0] = 0$
			$q[1] = 0$
			$q[2] = 0$
			$q[3] = a$

So let us see how Vivado optimized your full adder. Go back to your calculator project and select one of the 'Full Adder' modules in the design sources and set it as the 'Top Module'. (All other instantiated 'Full Adder' modules will then also become top modules; see below.)

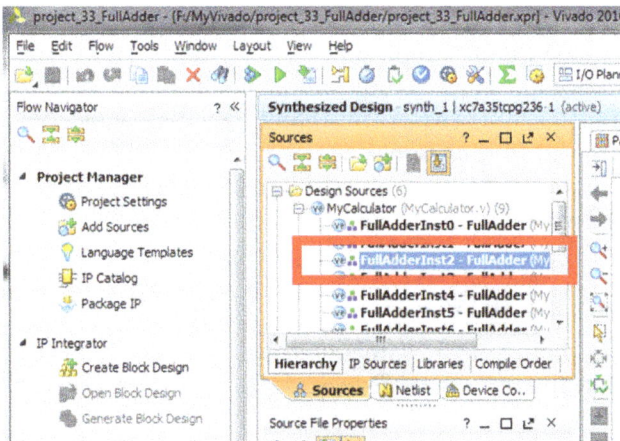

Synthesize your design and click on 'Synthesis/Synthesized Design/Schematic' (box 1 in the screenshot below). You will observe that the Xilinx compiler synthesized your Boolean logic design for the carry and sum bit using two LUTs (2). A straightforward way to implement a LUT is to use a MUX circuit with its $a[i]$ inputs set to the (constant) values in the LUT. In this example, the Boolean logic for the carry and adder bit has been replaced by two 8-to-1 MUX circuits connected to their corresponding LUT. (Inputs $I0$ through $I2$ are the MUX's selector lines.) It might be interesting to see what the actual input values, i.e. the LUT values, are. Click on the carry LUT (2) and then open the 'Truth table' tab (3) under 'Cell Properties'. In addition to displaying the truth table (4), which should look identical to the one you worked out earlier, you will also see (at the top) the Boolean expression that it implements.

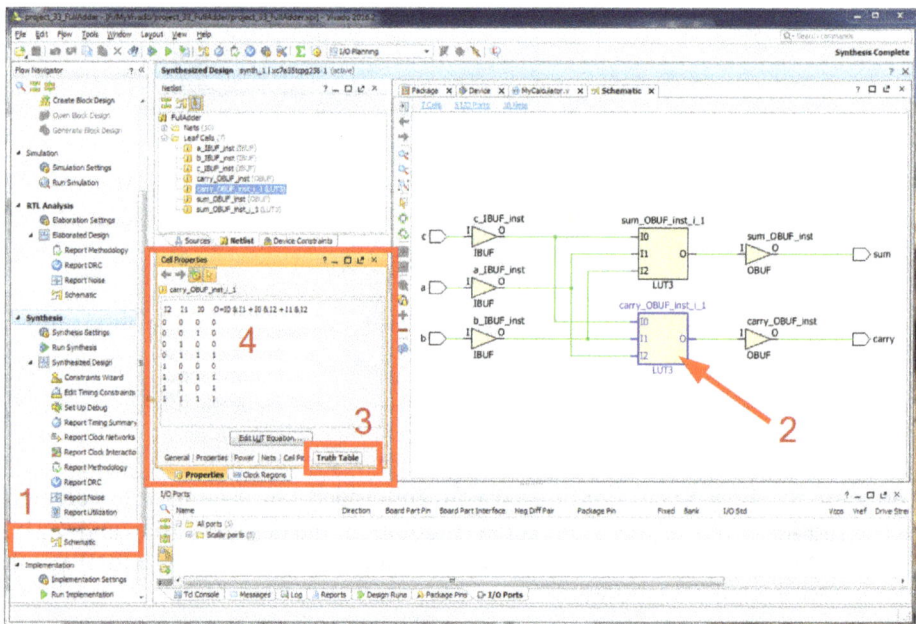

This example shows that Vivado prefers LUTs and MUXs over Boolean logic. We do not have to understand exactly why that is the case, sufficient to say that it solves many timing delay issues. At this point, you may wonder how far you may take the process of substituting Boolean logic gates with LUTs and MUXs. You will find the answer in the next few pages after you learn how to implement a MUX in hardware and software.

3.3.2 Review exercises: MUX/DEMUX

1. Before we show you how to implement the MUX in Verilog, write the complete truth table for the 2-to-1 MUX. (An abbreviated version is shown

in table 3.4.) Find its Boolean logic; draw its circuit using basic Boolean gates.

2. Extend your design to a 4-to-1 MUX. Do you see any pattern as to how to extend it to an even larger MUX, for example an 8-to-1 MUX?
3. For a 2^n-to-1 MUX, how many selector bits do you need?

3.3.3 Verilog implementation: conditional logic with a single logic condition

Using combinational Verilog logic, a 2-to-1 MUX is implemented by:

```
assign Wire Name = ( Logic Condition ) ? Value True : Value False;
```

For example, a complete module for the 2-to-1 MUX, shown above, is

```
module mux2(
    input [1:0] a,
    input sel,
    output q
    );

    assign q = (sel == 1'b0 ) ? a[0] : a[1];

endmodule
```

It works as follows: if the logic condition (sel == 0) is true, $a[0]$ will be assigned to q, else, $a[1]$ will be assigned to q. (Note the new relational operator '==' which is often confused with the assignment '=' operator which has a totally different purpose.)

Table 3.5. List of Verilog relational operators.

==	Is equal?
!=	Is not equal?
>	Is greater than?
>=	Is greater than or equal to?
<	Is less than?
<=	Is less than or equal to?

Table 3.5 is a table of Verilog relational and equality operators that you may use to determine the logic condition.

3.3.4 Verilog implementation: conditional logic with a multiple logic condition

In the previous case, the 2-to-1 MUX had only two input conditions which could be handled by a single logic condition similar to an 'if true do V else do W'. However, when working with a 4-to-1 (or even larger) MUX, multiple input conditions exist.

Of course, at any given time, only one of them can be true but every one of them has to be evaluated. In that case the conditions expand to:

if A is true do V,
        else if B is true do W,
        else if C is true do X,
        else do Y.

In Verilog, this is expressed as:

```
assign Wire Name = ( Logic Condition A ) ? Value V True :
    ( Logic Condition B ) ? Value W True :
    ( Logic Condition C ) ? Value X True :
     Y Value;   //Y Value is default case when none of the others are
correct
```

Of course, this syntax can be expanded further to any number of logic conditions. Going back to the 4-to-1 MUX shown earlier, its Verilog implementation is

```
module mux4(
    input [3:0] a,
    input [1:0] sel,
    output q
    );

    assign q = (sel == 2'b00) ? a[0] :
               (sel == 2'b01) ? a[1] :
               (sel == 2'b10) ? a[2] :
                a[3] ;  // default case if all others do not apply
endmodule
```

3.3.5 Review exercise: MUX/DEMUX II

1. Write a complete Verilog module for 1-to-2 DEMUX. Assume that it has an input named a and a 2-bit output vector named q.

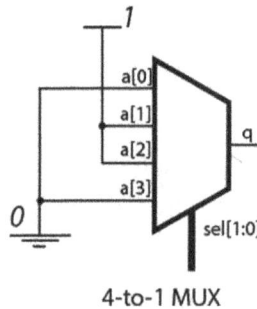

4-to-1 MUX

Figure 3.10. Schematic of a 4-to-1 multiplexer with fixed inputs, a[0] through to a[3].

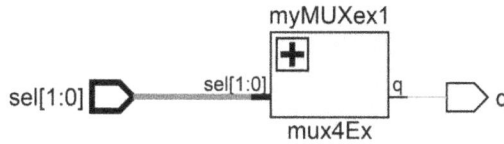

Figure 3.11. Schematic elaboration of the 4-to-1 MUX module shown in figure 3.10. Since a[0] through to a[3] never change, they have been implemented inside the module as internal wires with constant values; hence, they are not shown as inputs.

3.3.6 Verilog project: look-up table (encoder)

Figure 3.10 is a simple 4-to-1 MUX with a fixed or constant LUT. It will be your job to implement it in a Verilog project and study the behavior of the output q as a function of its selector lines sel[1:0].

Implement it in a completely new Verilog project with a module similar to the one shown in figure 3.11 and assign sel[1:0] to switches SW0 and SW1 and q to LED0. (Since the $a[i]$ inputs to the MUX are constant, your module will only have one input (vector) and one output, as shown in figure 3.11.)

Test your Verilog circuit and create the truth table for all combinations of SW0 and SW1. Its truth table should look familiar to you because it mimics a simple Boolean gate. Which one?

3.3.7 Conclusion on MUXs and FPGAs

As you saw in the preceding exercise, a combinational logic circuit, no matter how simple or complex, can be implemented entirely with LUTs and MUXs. The implication is profound: if you replace a (fundamental) logic gate with a program-mable LUT and a MUX you could create a 'generic' logic gate. For example, if you wanted your generic gate to behave like an AND gate, you would program its LUT for an AND gate. When you change your mind, you could reprogram it and change it into an NAND, XOR gate, etc, the sky is the limit. Next imagine you had a large collection of these generic gates, say a few million, all connected with a large network of MUXs. If such a device existed, you could 'on the fly' program it and implement any logic design you choose, from a simple XOR gate to an entire microprocessor with affiliated communication ports. As you may have guessed, such devices do exist and they are called FPGAs and that is what you have been working with for the last 2 chapters. There is only one more component that real FPGAs incorporate and that is memory. We will have more to say about that in the next two chapters.

3.3.8 Review exercises: MUX/DEMUX III

1. Create your Verilog code and elaborated schematic for the LUT (encoder) example above. What basic Boolean gate does it simulate? (This is the third time you have built this particular gate, each time using a completely different approach!)

Chapter 4

FPGA and VERILOG part II: sequential logic

Additional reading

Read the following sections (pages) from Horowitz P and Hill W 2015 *The Art of Electronics* 3rd edn (New York: Cambridge University Press) for the sections of this book indicated:

- Section 10.4 (pp 728–32) for section 4.1.
- Section 10.5 (pp 740–4) for section 4.3.
- Section 13.2.8 (pp 879–88) for section 4.5.

In this and the following chapters we will add sequential logic to our combinational digital circuits. The purpose of this is to give the circuits memory. We will apply this feature and create counters and a digital-to-analog converter capable of playing a short piece of music.

4.1 Sequential logic: latches

Before we start implementing the sequential logic circuit, let us first define what makes a circuit combinational (sometimes also called combinatorial) or sequential. Consider the black box shown in figure 4.1, which has two inputs, A and B, and an output Q.

The key question then is: can we determine Q entirely from the current state of A and B or do we need to know the circuit's previous history?

For example, for a combinational circuit such as a simple AND gate, Q is uniquely defined for any value of A and B, independent of its previous history. However, if we place a sequential logic device in the black box, knowing A and B alone can be insufficient in predicting the state of Q. The reason is that sequential devices retain their states even if one of the inputs changes. (Usually, one of the inputs of such a device is special and it alone controls whether or not the output remains unchanged, independent of the state of the other input.)

doi:10.1088/978-1-6817-4660-9ch4
4-1

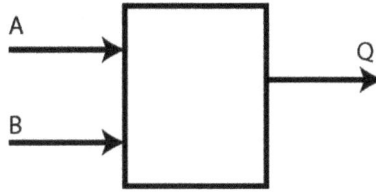

Figure 4.1. A 'black-box' module with two inputs and one output.

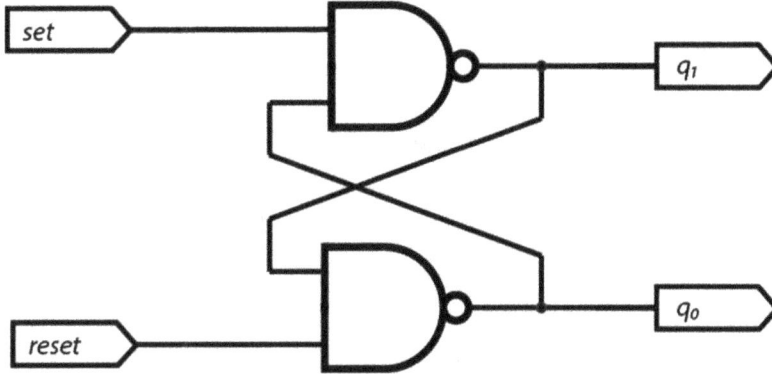

Figure 4.2. An SR latch.

Table 4.1. Transparent latch truth table.

Case	set	reset	$q0$	$q1$
i	0	1		
i	1	0		
ii	0	0		
iii	1	1		
iii	1	1		

What makes the sequential versus combinational logic distinction a bit more puzzling is that when you examine the sequential logic circuit at the gate level you will find it is composed exclusively of the familiar combinational logic elements, such as NAND or NOR gates. An example of such a sequential circuit, called an 'SR latch', is shown in figure 4.2. Since it consists of only two NAND gates it is tempting to treat it as a combinational circuit. However, by cleverly interconnecting the outputs, i.e. by providing feedback, the amnesic combinational circuit now exhibits hysteresis and it has memory. For certain input states, its output state can only be determined by knowing its previous output state(s).

To understand how the transparent latch works, complete the truth table in table 4.1.

This truth table can be grouped into three distinct cases: for cases i and ii, its outputs are uniquely defined independent of their previous history. However, for case iii, two viable output state options exist, both completely in agreement with Boolean logic rules. So which state wins? In case iii, the status quo prevails and its output will not change as long as $q0 = \sim q1$. (Since case ii does not agree with either of the case iii states, it is impossible to predict what state the outputs will be if case iii has been applied directly after case ii. Therefore, case ii should be avoided!)

From case iii, it follows then that an SR latch's output state can no longer be predicted solely from its inputs. Instead, we also have to consider its previous history. To determine its unique output state we must also know which specific case i preceded the current case iii.

4.1.1 Verilog implementation of a transparent latch

The SR latch circuit above can be implemented in hardware using 7400 ASIC logic NAND gates. (If you are curious try it.) However, a direct translation to Verilog fails during compilation because of 'race conditions', meaning the compiler does not know how to handle components where the output is instantaneously applied back to the input.

Instead, we need to implement a similar latch, called a transparent latch, using the Verilog construct previously encountered for a multiplexer:

```
assign q = set ? q : a;
```

Although the statement appears simple, it is rather tricky. Try to understand it and predict its behavior before you implement the simple Verilog module below, which uses a similar statement. Create a new project; add a constraint file and uncomment the lines pertaining to button 'btnC', the switches 'sw' and the LEDs 'led'.

```
module MyLatch(
    input btnC,     // Center button: pressed is "high", released, i.e., default is "low"
    input [15:0] sw,
    output [15:0] led
    );

    wire set;
    assign set = btnC;

    assign led = set?led:sw;

endmodule
```

Generate the bit-file and program your BASYS3 board.

Now test your latch:

1. Press the center button, BTNC, which corresponds to *high*; move the switches SW0 *to* SW15 and observe the corresponding LEDs LD0 to LD15.
2. Release the center button and again move the switches. Observe the state of the corresponding LEDs.
3. Go back to step 1.

4.1.2 Review exercises

1. You may have previously encountered the concept of converting an analog circuit without memory into one with memory when working with operational amplifiers (op-amps). Are there any similarities with the simple transparent latch which relies on only combinational logic?

2. Complete the truth table for the SR latch, i.e. table 4.1.

3. Describe, in a few sentences, the behavior of the latch module 'MyLatch' shown above when 'set' is high and when 'set' is low. How do the words 'transparent' and 'latch' relate to its behavior?

4. Complete the timing diagram in figure 4.3 for the following Verilog statement:

```
assign A = SET ? A:B;
```

5. The circuit in figure 4.4. uses a mechanical switch to control the outputs V_x and V_y. Theoretically, one would expect the following results as the switch is moved through positions a, b and c: when the switch makes contact in position a, $V_x = V_{on}$ and $V_y = 0$ V. Once the switch leaves position a, i.e. 'breaks' contact, and is between the contacts, position b, $V_y = V_x = V_{on}$. Finally, when it again makes contact at c, $V_x = 0$ and $V_y = V_{on}$. However, real mechanical switches and relays often bounce off multiple times at first impact at the contact point c. Note that switches do not bounce all the way back from c to a; instead they just keep bouncing between b and c until they ultimately settle down at c. Typically the break at a is clean. While this 'bouncing' effect is real and often can be ignored, it has a detrimental effect on any circuit attached to V_x counting the number of times the switch has been moved from a to c because each bounce adds a false count. A simple cure for this phenomenon is attaching a debouncing circuit in the form of the SR latch shown in figure 4.2. Specifically, draw a timing diagram that shows V_x and V_y as a function of time when V_x has been connected to set and V_y to the reset input. Assume that the switch initially rests in position a and then breaks from it. See figure 4.4.

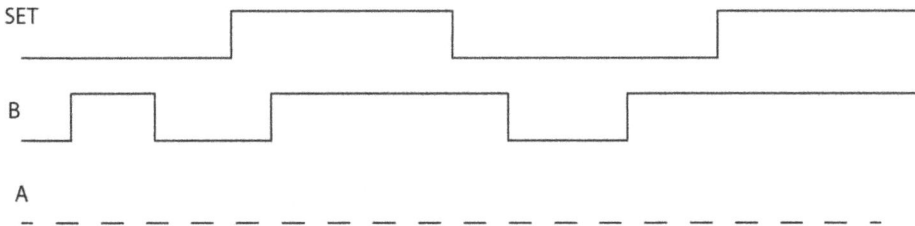

Figure 4.3. Complete the timing diagram for the circuit shown in figure 4.4. when the inputs 'SET' and 'B' are in the state indicated above.

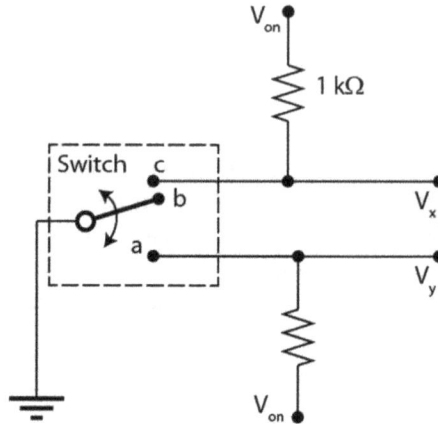

Figure 4.4. A switch with an RS-latch to suppress the 'bouncing' effect when the wiper, b, makes contact at either position a or c.

Figure 4.5. D-type flip-flop.

4.2 Sequential logic: flip-flops

By far the most common sequential logic circuit element is the flip-flop, which you will start using in this exercise. It contains multiple latches in series so that at any one time only one of them is active. This arrangement is chosen to ensure that the flip-flop is set, or 'loaded', during an infinitesimal time interval when its 'clock' input changes logic levels.

Flip-flops come in a few different families, with the D-type, shown in figure 4.5, being the most prevalent.

Of the two flip-flop inputs, the clock input, CLK, controls the precise moment when the D-input's state will be latched and stored inside the flip-flop. (The clock input is usually drawn with a small triangle to mark its unique purpose.) Depending on the type, flip-flops latch the D-input on the active clock edge which can either be on the positive clock transition (figure 4.5, left panel) or on the negative transition (right panel).

After a brief propagation delay, the internally latched input state appears at its output Q. At the same time, its complementary output, i.e. its negated input, appears at $\sim Q$. Until the next active clock edge transition occurs, the flip-flop's outputs will remain fixed and its outputs are not affected by any changes at its D-type input.

As a simple analogy consider a 1-bit camera: the D-input corresponds to the camera's lens and the clock input is the shutter release button. Although the lens may be facing any number of different subjects, the camera only takes a 1-bit picture when the shutter release button is activated. Specifically, it will only take a picture either when you release the button (positive edge triggered) or when you press it down (negative edge triggered). *Important*: it is never the case that the camera takes two consecutive pictures, i.e. one when the shutter is pressed and one when it is released. It will only activate once, either when the shutter is pressed or when it is released. After a short processing time, the (1-bit) image will appear at its display, the output Q, and its inverse image at $\sim Q$. It will remain stored until you override it by taking a new picture when you activate the shutter release button again.

While the circuit elements inside the flip-flop are entirely combinational, the flip-flop itself is a sequential logic element. Its main purpose is to store (a 1-bit) state in its memory and its output depends on the state of the D-input at the preceding active clock edge transition.

4.2.1 Verilog implementation of a D-type flip-flop

The Verilog implementation of a D-Type flip-flop requires new language components to accurately reflect the time-dependent nature of this sequential logic element. While the combinational logic was completely defined by gates and wires, sequential logic uses *registers*. You may consider a register to be a 1-bit memory cell, i.e. the *output* of a D-type flip-flop.

Also, while the state of a wire changes the instant its assigned (input) state changes, the output of a register is only affected during the logic transition of its associated clock input. This association of a clock signal with a register is made in Verilog by the 'always-block' statement. An example is shown in the code snippet below which implements a simple positive edge triggered D-type flip-flop, similar to the one shown in figure 4.5:

```
reg Q = 0;
always@( posedge CLK) begin
     Q <= D;
end
```

Let us examine each line of this code, staring with

```
reg Q = 0;
```

Each register must be explicitly declared with the Verilog keyword 'reg'. Registers may exist as internal registers, like the one named 'Q' in the code snippet above, or they can be a module's output ports as shown in the example below:

```
module MySimpleDTypeFF(
    input CLK,
    input D,
    output reg Q = 0
    );

    always@(posedge CLK) begin
        Q <= D;
        end

endmodule
```

(*Note*: while wires can be declared as input or output ports, registers can only be declared as output ports. The reason being that registers refer to the output of the flip-flop and *never* to its inputs!)

Unlike wires, registers can and should be initialized. (Remember, initializing wires makes little sense unless you want them to be constants.) Initializing a register is equivalent to setting or resetting a flip-flop at the powering-up of the circuit.

Finally, similar to wires, an entire collection of registers can be declared as a register vector. For example, declaring a 16-bit internal vector named 'goldy' uses the following syntax: 'reg [15:0] goldy;'. The notation you are already familiar with for concatenating and splitting vectors is identical for wire and register vectors.

Let us look at the second line of code:

```
always@( posedge CLK) begin
```

All assignments to a specific register must be made within a single always-block statement. It specifies the clock source(s) and the active edge type, either 'posedge' or 'negedge'. The assignments are enclosed within 'begin' and 'end' keywords. (These keywords are optional if they enclose only one assignment; for example, they could be omitted in the example shown above.)

The third line of code is shown below:

```
Q <= D;
```

Unlike wires which use the Verilog 'assign' keyword, registers do not use the 'assign' keyword. Instead they use the non-blocking assignment operator, <=, as shown above.

The 'MySimpleDTypeFF' module shown above lacks the complementary $\sim Q$ output that a typical D-type flip-flop provides. This functionality can be implemented two different ways using either sequential or combinational logic. Applying sequential logic and declaring a second register named 'NQ' is shown below:

```
module MyDTypeFFV1(
    input CLK,
    input D,
    output reg Q = 0,
    output reg NQ = 1
    );

    always@(posedge CLK) begin
        Q <= D;
        NQ <= ~D;
        end

    endmodule
```

A functionally equivalent version using both combinational and sequential logic is shown below:

```
module MyDTypeFFV2(
    input CLK,
    input D,
    output reg Q = 0,
    output NQ
    );

    always@(posedge CLK) begin
        Q <= D;
        end

    assign NQ = ~Q;
endmodule
```

Since combinational statements are not directly controlled by a clock you should not place them inside an always-block. Instead, see the example above where the combinational statement, 'assign NQ = ~Q;' was placed outside the always-block for exactly this reason. (Of course, registers must always be placed inside an always-block since they are controlled by the clock.)

4.2.2 Flip-flop exercise 1

Shown below is the slightly modified version of the 'MyLatch' module used in the previous section. (Changes are shown in bold face.) Instead of the transparent latches, the version below uses a 16-bit register vector named 'led' to implement 16 D-type flip-flops which are controlled by button BTNC and the switches. See the elaborated design schematic in figure 4.6.

```
module MyFF(
    input btnC,  // Note pressed is HI, released, i.e., default is LO
    input [15:0] sw,
    output reg [15:0] led
    );

    wire clock;
    assign clock = btnC;

    always@(posedge clock) begin
        led <= sw;
      end

endmodule
```

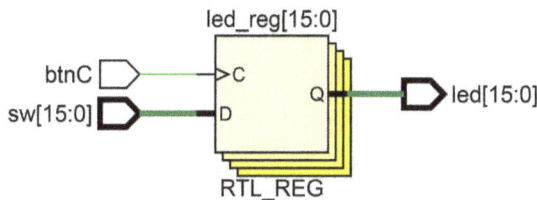

Figure 4.6. Elaborated schematic of 16 D-type flip-flops to store the individual switch settings when button 'btnC' is pressed.

Modify your 'MyLatch' module from the previous section to reflect the changes shown below. Note that when you will try to generate its bit-file you will get a 'Place-Design' error message:

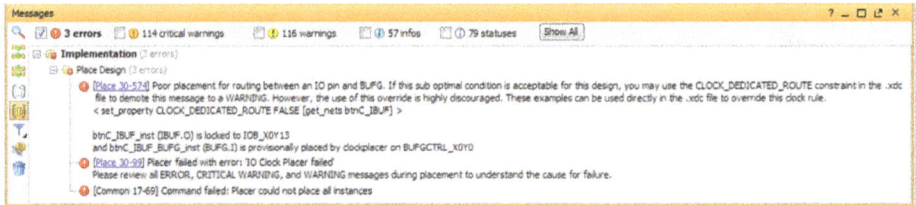

This error message is caused by our choice to assign button 'btnC' for the clock signal to the flip-flops. As we have already pointed out, the clock input to a flip-flop is a very special signal. Since the clock input has to be able to process signals into hundreds of megahertz, specific 'clock pins' have been designated on the FPGA. Unfortunately, none of the buttons or switches on the BASYS3 board have been connected to these reserved clock pins. (The reason being that, as you have seen in the previous section, mechanical switches generally suffer from 'bouncing' and therefore should only be connected after proper debouncing.)

A simple work-around the pin assignment problem exists by overriding the compiler warning message. Essentially, we are telling the compiler that we are aware of what we are doing and we take responsibility for assigning signals to non-designated clock pins of the FPGA.

Open your project's constraint file and add the line shown below in bold directly below the 'btnC' pin assignments:

```
##Buttons

set_property PACKAGE_PIN U18 [get_ports btnC]
    set_property IOSTANDARD LVCMOS33 [get_ports btnC]
    set_property CLOCK_DEDICATED_ROUTE FALSE [get_nets btnC_IBUF]
```

Repeat the previous testing procedure for the transparent latch and describe, in a few sentences, the behavior for the D-type flip-flop used in this exercise:

1. Press the center button BTNC, which corresponds to *high*; move the switches SW0 to SW15 and observe the corresponding LEDs LD0 to LD15.
2. Release the center button and again move the switches. Observe the state of the corresponding LEDs.
3. Go back to step 1.

In a few sentences, in plain English, describe the behavior of the flip-flop module when 'set' is high and when 'set' is low. How is it similar to and different from the latch in review exercise 4.1.2, question 3?

4.2.3 Flip-flop exercise 2

Now observe this extremely simple, yet very fundamental and useful flip-flop arrangement shown below, where the negated output is fed back to its input:

```
module MyFFV1(
    input btnC,  // Note pressed is HI, released, i.e., default is LO
//    input [15:0] sw,  //no longer needed
    output reg [15:0] led
    );

    wire clock;
    assign clock = btnC;

    always@(posedge clock) begin
        led <= ~led;
    end

endmodule
```

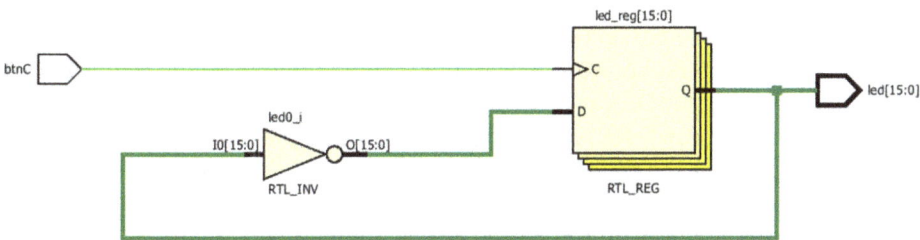

Figure 4.7. In this elaborated schematic, the negated flip-flop outputs are fed back into the flip-flops.

Modify the module from the previous exercise and then test the behavior of the one shown above by repeatedly pressing BTNC. (*Note*: the switches are no longer needed and have been disabled.) Do the LEDs light up?

4.2.4 Review exercises

1. Complete the timing diagram shown below for a positive edge clocked D-type flip-flop and show its output Q and $\sim Q$. Do the same for a negative edge clocked flip-flop. Describe any assumptions you made in the timing diagram.

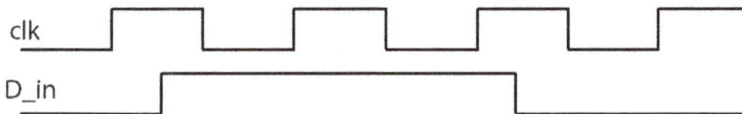

2. In a few sentences, in plain English, describe the behavior of the flip-flop module in section 4.2.2, flip-flop exercise 1, when 'set' is high and when 'set' is low. How is it similar to and different from the latch in exercise 4.1.2, question 3. Is it still 'transparent'?
3. Make a timing diagram for the module in flip-flop exercise 2, section 4.2.3. Specifically, show BTNC and one of the LEDs as a function of time. For

simplicity, assume BTNC is pressed at regular time intervals, similar to the clock signal shown in the timing diagram of the first exercise in this section.

4. If BTNC in flip-flop exercise 2 was pressed N times then how many times would the LEDs light up? If the button was operated at regular intervals at a frequency f_0 then at what frequency would the LEDs blink?

4.3 Sequential logic: fundamental counters

In the last exercise you connected the negated flip-flop output to the D-input. Each time button BTNC was pressed, the flip-flop's output 'toggled' between high and low, or 0 to 1. This sequence of numbers also corresponds to be the output of the least significant bit of a binary counter, i.e. 0, 1, 10, 11, 100. In other words, by connecting the negative output of the flip-flop to back to its input, you surreptitiously built a 1-bit counter.

This 1-bit counter design can be expanded into a multi-bit counter by cascading one of its outputs to the clock input of the next highest 1-bit counter. (See the diagram below for one version of how to implement a multi-bit counter.) The $Q0$ output of the first flip-flop corresponds to the LSB; the output of the last flip-flop in the cascade, $Q3$, corresponds to the MSB.

Since the clock input for each subsequent flip-flop will be delayed by the propagation delays of the previous flip-flops, this type of counter is called a 'ripple-counter'. (The clock signal 'ripples' down the chain from the LSB to the MSB.) When implementing this circuit in an FPGA the propagation delays usually are not an issue and for practical purposes we can assume that they are infinitely small.

We have not yet determined whether the counter shown in the diagram in figure 4.8 will count up or down. This will be your job to figure out in the exercise below.

4.3.1 Counter exercise 1

Let us implement the 4-bit ripple counter schematic shown above in a new project and figure out if it counts up or down.

Similar to the previous exercise, use 'btnC' as the input clock. (Do not forget to include again the 'set_property' 'CLOCK_DEDICATED_ROUTE FALSE [get_nets btnC_IBUF]' statement in the constraint file.)

Instead of 'hard coding' the four D-type flip-flops in your project, instantiate them four times. In the previous section, you were given two different versions of such modules, i.e. 'MyDTypeFFV1' or 'MyDTypeFFV2'. Use the one you prefer and instantiate it four times as shown below. (You may want to declare internal 4-bit

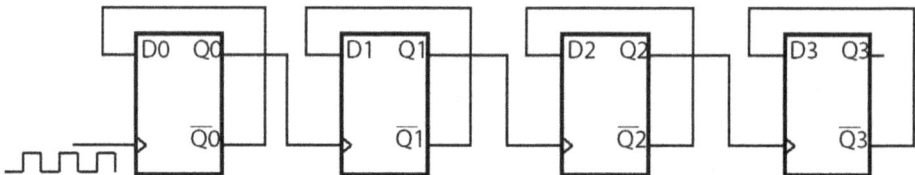

Figure 4.8. Schematic representation of a 4-bit ripple counter.

wire vectors Q and NQ for the values for $Q0$ through $Q3$ and $\sim Q0$ through $\sim Q3$, respectively.)

Finally, add the 'HexDisplayV2' module to your project and instantiate it to display the counter's Q output values. (If you need a refresher, review section 3.1.) Set the second and third argument of the 'HexDisplay' instantiation, the 'BCD_enable' and the 'Display_Enable' permanently to 1 as shown in the sample code below:

```
HexDisplayV2  myHexDisplay( clk, Q, 1, 1, seg, an );
```

Do not forget to add the system clock, 'clk', and the hex display ports 'seg' and 'an' to your top module and to uncomment the corresponding lines in the constraint file. The complete elaborated design schematic is shown in figure 4.9.

Press button BTNC and watch your counter counting. Is it counting up or down?

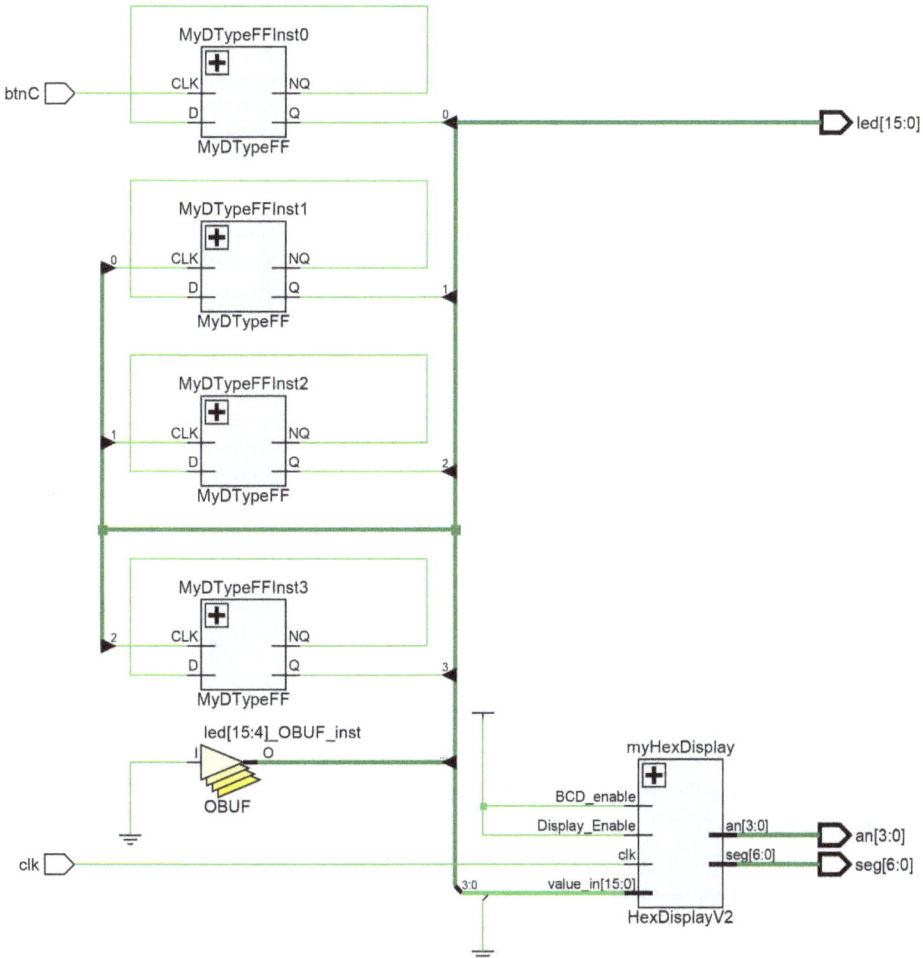

Figure 4.9. Elaborated design schematic of the 4-bit ripple counter and the hex display module.

4.3.2 Review exercises

1. Create the complete Verilog source code for the counter shown above. Does it count up or down?
2. Make a complete timing diagram for your counter. Specifically show your clock and $Q[0]$, $Q[1]$, $Q[2]$ and $Q[3]$. At the bottom of your timing diagram, clearly indicate the decimal value represented by Q at each state. Press BTNC and verify the timing diagram with the blinking LEDs on the BASYS3 board.
3. A binary counter can also be used as a frequency divider. As such, it is a very convenient and quick way to derive a slower (clock) signal from a faster one. When you look at your timing diagram from the previous question, what is the relationship between the frequency of the nth counter bit, f_n, to the original clock signal, f_0, driving the very first flip-flop? (*Hint*: see also your answer from review exercise 4.2.4, question 4.)
4. As you press button BTNC, you should notice that occasionally the counter seems to skip a count or two. (If you do not believe it, close your eyes and push the button n times; compare your count with the display. Which one is right?) How can you explain this behavior or what can you blame it on?

4.3.3 Counter exercise 2

Modify the counter from the previous exercise so that it will count the opposite way. In other words, if it was an up counter turn it into a down counter and vice versa. The modifications required are all minor. You may want to explore different active clock edges for the flip-flops, looking at the complementary outputs of the flip-flops or connecting the clock inputs to different outputs. There are at least three ways to accomplish this. Be creative!

4.3.4 Review exercises

1. Create the complete Verilog source code and elaborated design schematic for your counter. Does it now count up or down?
2. Make another timing diagram for your counter. Specifically show your clock and $Q[0]$, $Q[1]$, $Q[2]$ and $Q[3]$. At the bottom of your timing diagram, clearly indicate the decimal value represented by Q at each state. Press BTNC and verify the timing diagram with the blinking LEDs on the BASYS3 board.

4.4 Sequential logic: counters with logic conditions

In the previous section you implemented counters from first principles using nothing more than cascaded flip-flops and their complementary outputs. Since counters are fundamental to digital designs, the Verilog language contains a more efficient way to implement them, shown below. This syntax uses the addition operator for incrementing and the subtraction operator for decrementing a register vector. The example below increments (an n-bit register vector) x by one at each positive clock edge:

```
always@( posedge clk) begin
    x <= x + 1;                  // x is an n-bit register vector
    end                          //declared earlier
```

The maximum numerical value which can be expressed by an (unsigned) n-bit vector is $2^n - 1$. Incrementing (by one) past this value results in the counter 'rolling over' back to 0. (Similarly, decrementing a counter, which is at 0, by one results in it being set to $2^n - 1$.)

4.4.1 Example 1: previous exercise using adders

Applying the new syntax, we could have implemented the previous design, the 4-bit counter, with the code shown below in bold face. As you can see, there is no longer a need to instantiate the four 'MyDTypeFF' modules:

```
module MyCounterV3(
    input clk,
    input btnC,   // Note pressed is HI, released, i.e., default is LO
    output [15:0] led,

    output [6:0] seg,        //each bit corresponds to one of the 7 segments on the display
    output [3:0] an          //specifies which of the 4 displays is to be turned on
    );

    wire clock;
    assign clock = btnC;

    reg [3:0] my_counter = 0;    //4-bit register vector

    always@(posedge clock) begin  //4-bit counter
        my_counter <= my_counter+1;
        end

    assign led = my_counter;

    HexDisplayV2 myHexDisplay(  clk, my_counter, 1'b1, 1'b1, seg,    an    );

endmodule
```

Its corresponding elaborated design schematic is shown in figure 4.10, revealing the use of a mathematical adder and the 4-bit register vector, 'my_counter_reg'.

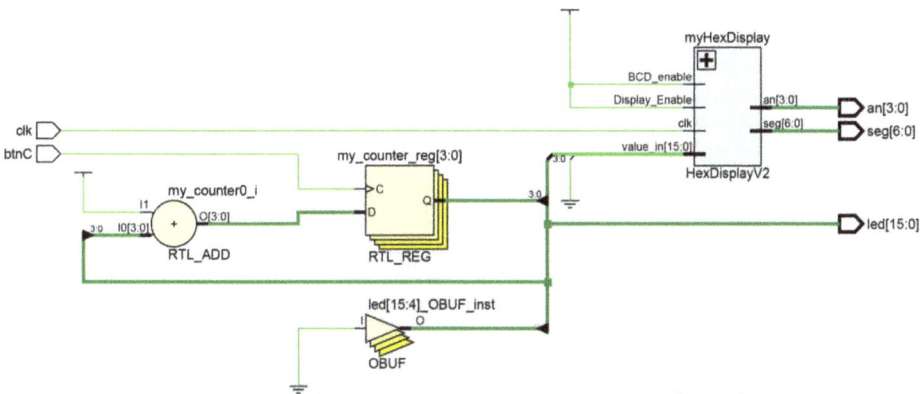

Figure 4.10. Elaborated design schematic of a 4-bit counter utilizing an adder to increment the counter registers.

4.4.2 Example 2: decade counter; sequential logic if–else statements

As we have already explained, a free running n-bit binary up-counter will roll over to 0 after it has reached its maximum value, 2^n-1. On the other hand, a decade counter resets itself after ten cycles. It counts from 0 up to 9 and then returns to 0. In other words, it never reaches 10, or for that matter, any larger value.

Such a counter module is useful. For example, if we were to cascade multiple instances together (as shown in figure 4.11) we have a counter that counts in decimal, or rather in binary coded decimals (BCD), instead of binary. Hence, we no longer would need a binary to decimal conversion module.

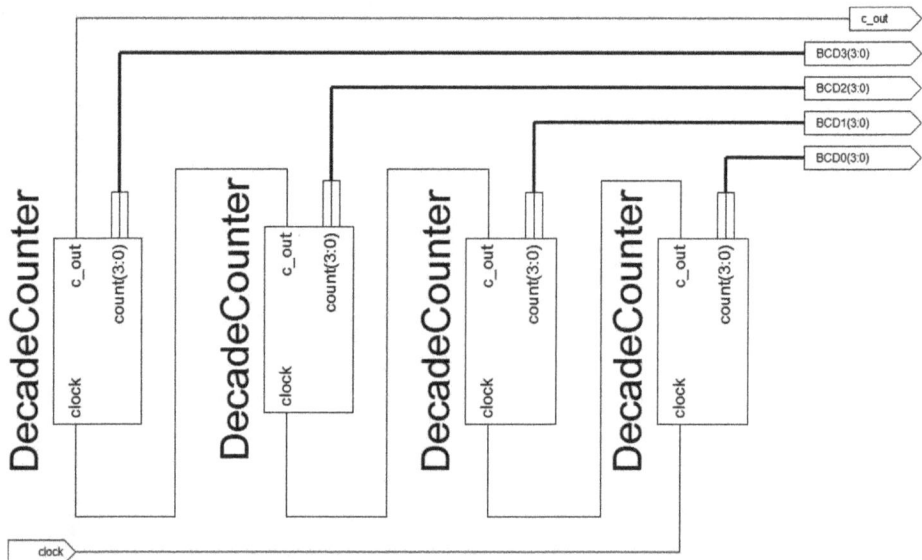

Figure 4.11. Block diagram of a 4-digit (decimal) decade counter with a range from 0000 to 9999.

To implement such a decade counter we need a method to force the free running binary counter to roll over to 0 once it has reached 9. In sequential logic, this can be accomplished by an *if–else* statement as shown in the example below:

```
reg [3:0] my_counter = 0;    //4-bit register vector

always@(posedge some_clock) begin
        if( my_counter == 9 ) begin
            my_counter <= 0;    //executed if above logic condition is true
            end
        else begin
            my_counter <= my_counter+1;  //executed for all other cases
            end
        end
```

Important: a very common mistake is to confuse the MUX operators, as in 'assign x = a ? b: c;' with the if–else statements shown above. Note that the MUX operators

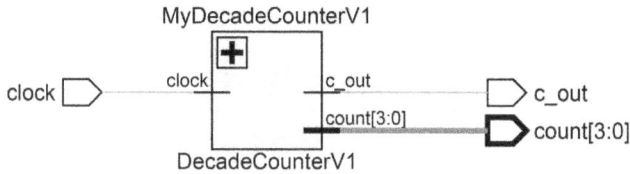

Figure 4.12. Elaborated design schematic of a single digit decade counter.

can only be applied to combinational logic components while the if–else operation only works with sequential logic. In other words, you should only use the if–else statements inside an always-block.

Let us implement the complete (single) decade counter module shown in figure 4.12. Its 4-bit 'count' output represents the current BCD count value, which is incremented at the rising edge of the 'clock' input. The 'c_out' output indicates that the counter has rolled over from 9 to 0; it remains high whenever the counter is 0 and low for all the other values. This output serves as the clock input for the next higher digit, as shown in the cascaded BCD counter design shown previously.

Two different versions of such a decade counter module are shown below. They differ in their implementation of the 'c_out' output: the first one uses a combinational approach and the second a sequential one. Please study the two examples.

```
module DecadeCounterV1(
    input clock,
    output reg [3:0] count = 0,
    output c_out
    );

    always@(posedge clock)begin
            if (count == 9) begin
                    count <= 0;
                    end
            else begin
                    count <= count +1;
                    end
            end

    assign c_out = (count==0) ? 1 : 0; //combinational logic
endmodule
```

```
module DecadeCounterV2(
    input clock,
    output reg [3:0] count = 0,
    output reg c_out = 0
    );

    always@(posedge clock) begin
            if (count == 9) begin
                    count <= 0;
                    c_out <= 1;    //sequential logic
                    end
            else begin
                    count <= count +1;
                    c_out <= 0;    //sequential logic
                    end
            end
endmodule
```

Can you spot a subtle problem with the first version? It only occurs right at startup when such modules are cascaded together as in figure 4.11.

Note: Drawing is NOT to Scale!

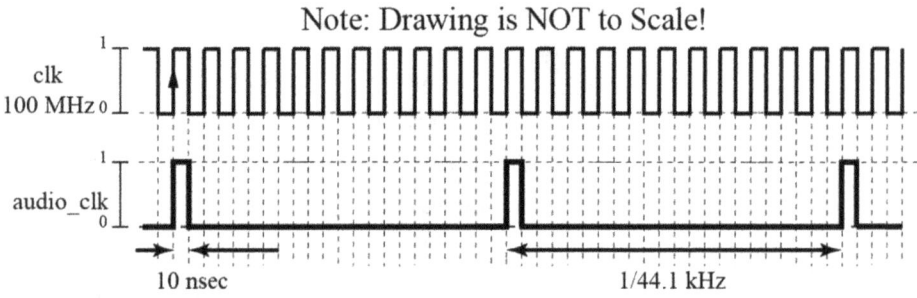

Figure 4.13. Timing diagram of the 'AudioClock' module shown directly below.

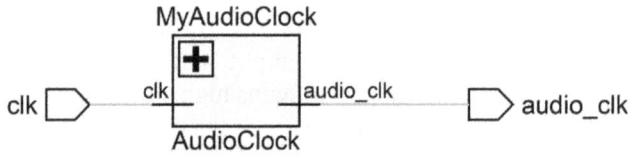

Figure 4.14. Elaborated design schematic of the AudioClock module which outputs a 10 ns pulse every 1/44.1 kHz, i.e., about every 23 microseconds.

4.4.3 Exercise 1: 44.1 kHz audio clock

For consumer electronics, audio data are typically updated at a rate of 44.1 kHz. This specific rate was selected because it can truthfully replicate audio signals up to half of this frequency and, therefore, it covers most of the audible range of humans. (The reason it can only replicate signals at half of its sampling frequency is due to Nyquist's theorem.)

The aim of this chapter is to build a simple audio player. In this exercise your task will be to build an 'audio-clock' module which outputs a 10 ns (high) pulse at the standard audio rate of 44.1 kHz.

You will use the existing 100 MHz system clock on the BASYS3 board to increment a counter. (You have used this system clock previously to control the Hex display module.) Similar to the decade counter example shown in figure 4.11, once your counter reaches a preset value, it will reset itself to 0 and send out a high signal for one system clock cycle and then repeat the cycle.

Figure out the value the counter needs to reach so that it resets itself 44 100 times a second if it is controlled by the system clock running at 100 MHz. (*Hint*: how many system clock cycles elapse in a 1/(44.1 kHz) period?)

Write the complete module, whose elaborated schematic is shown in figure 4.14.

4.4.4 Connecting signals to and from the BASYS3 board

Since this module is a crucial component of your audio player, you want to test it by sending a signal from the BASYS3 board directly to a test instrument. Ideally you want to observe the output signal on a (digital) oscilloscope. (You could also measure its frequency with a frequency counter.)

Figure 4.15. BASYS3 board with its JB Pmod input/output connector highlighted. (Reproduced with permission from *Basys3™ FPGA Board Reference Manual* (2016, Pullman, WA: Digilent). Copyright 2016 Digilent Inc.)

Figure 4.16. The Digilent BNC1 Pmod connector simplifies the interconnection of the BNC cables (from your scope) to the BASYS3 board. The blue jumpers select which of the four Pmod signals will be connected to its nearest BNC connector.

For testing, you will instantiate your existing audio module in a new top module and send its output to pin 1 of the JB Pmod connector, identified in figure 4.15.

Create the new top module with input port 'clk' and output ports [7:0] JB:

```
module AudioClockTop(
    input clk,
    output [7:0] JB
    );
```

Instantiate your original audio clock module in this top module and assign the 'audio_clk' signal to JB[0].

Add the constraint file to your project and uncomment the lines pertaining to the system clock, 'clk', and the JB inputs/outputs.

The most reliable way to connect a cable to the Pmod connectors is through the BNC1 Pmod connector sold by Digilent Inc, shown in figure 4.16. If you have one in your lab, plug it into the *upper* row of Pmod connector JB. Please check the position of the blue jumper on the BNC1 connector. It selects which of the four input pins, JB [0] through JB[3], is connected with its adjacent BNC connector. In position A, as shown above in figure 4.16, the BNC is connected to JB[0]; similarly, in position B, JB[1] will be connected, etc.

If you do not have such a connector, insert wires directly into the JB Pmod connector shown in figure 4.17. Pin 1 corresponds to JB[0] and pin 5 is ground.

Connect one end of a BNC cable to the BNC1 Pmod shown in figure 4.16 and the other end to channel 1 of a digital scope. To view these very brief pulses, you may need to adjust the time scale of the scope and you must ensure that the scope is being triggered properly.

When you zoom-in on your pulse, you may observe a rather messy signal with multiple pulses similar to the screenshot in figure 4.18. This effect is due to impedance mismatch of the cable and the scope and it causes the pulse to be reflected multiple times between the board and the scope. This can be avoided by adding a BNC-Tee and a 50 Ohm terminator at the scope input, as shown in figure 4.19.

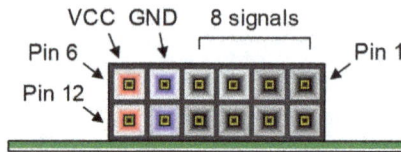

Figure 4.17. Pmod connectors; front view. (Reproduced with permission from *Basys3*™ *FPGA Board Reference Manual* (2016, Pullman, WA: Digilent). Copyright 2016 Digilent Inc.)

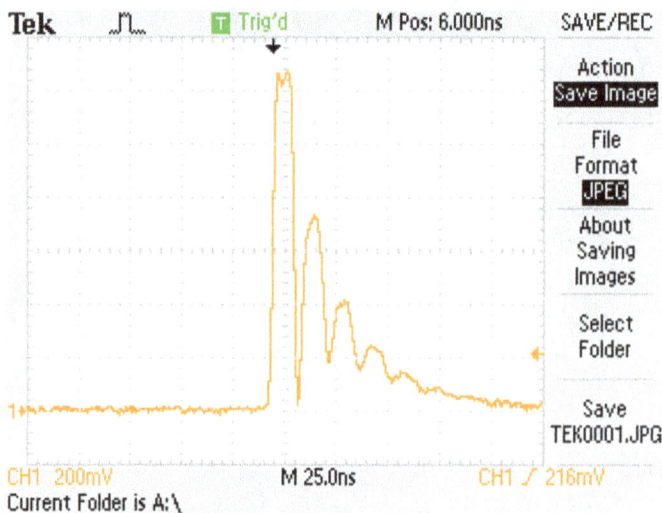

Figure 4.18. Noisy AudioClock output signal as a function of time. The reason for observing the 'ringing', or reflected pulses, is due to an impedance mismatch between the low output impedance of the board and the high input impedance of the scope.

Figure 4.19. A BNC Tee with a 50 Ohm terminator which will eliminate (most) of the ringing by effectively reducing the input impedance of the scope from Megohms down to 50 Ohms.

Program your board with the bit-file. If your code works correctly, you should observe a frequency that is within 15 Hz of the target 44.1 kHz rate. Take a screenshot of your signal from the scope.

4.4.5 Review exercises

1. Create the complete Verilog source code for your 44.1 kHz audio clock module.
2. What frequency did you measure for your audio clock module with the scope?
3. Include a screenshot of a single pulse of your signal from the scope after you added the 50 Ohm terminator.

4.5 Pulse width modulation (PWM)

4.5.1 Introduction: digital-to-analog conversion methods, pulse width modulation

The exercises in this section will guide you further towards the final project where you will be building the audio player. Our design is still missing a method to convert digital signal data (representing the amplitude of our audio sound at a specific time) into an analog signal which can be played on a speaker. This conversion of a digital signal into a corresponding analog signal is accomplished by using a programmable voltage supply, more typically called a digital-to-analog (D2A) converter. These devices are usually built with analog and digital circuit components. However, in this section we will explore a method called pulse width modulation (PWM) for D2A conversion and it only requires digital components which already exist on the BASYS3 boards. This technique benefits and suffers from the advantages and shortcomings of the D2A (and also in the analog-to-digital) conversion process and we will explore some of these characteristics.

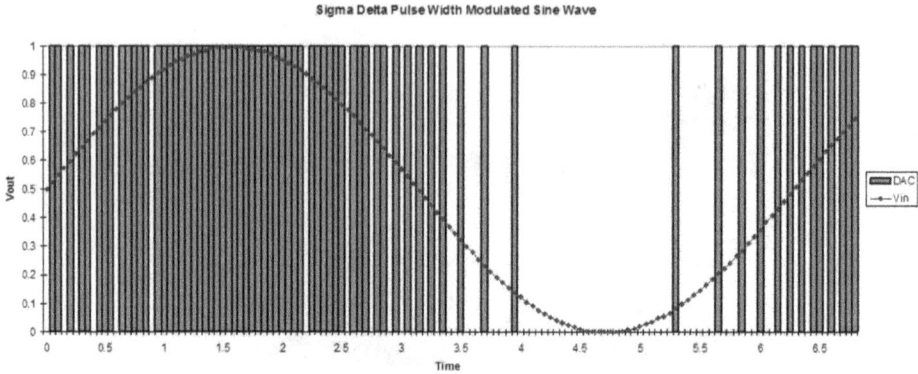

Figure 4.20. Graph of an analog sine wave and its (Sigma Delta) pulse width modulated equivalent output.

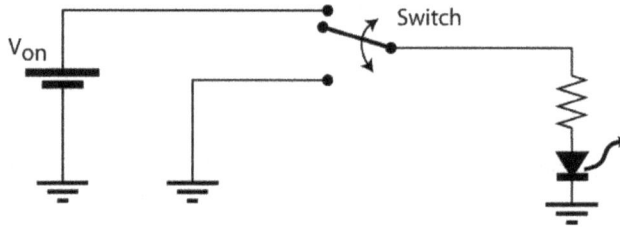

Figure 4.21. LED with manually operated switch.

4.5.2 Introduction: light (LED) dimmer

In this exercise you will build a simple PWM circuit to control the brightness of an LED. While the LED on a BASYS3 board can only be controlled through two digital voltage levels, i.e. high or low, you will explore the PWM technique to give the LED the appearance as if it were controlled by an almost continuously adjustable analog voltage. You will use switches SW0 through SW7 to adjust the intensity of the LEDs on the board.

To illustrate the concept of PWM, consider the circuit in figure 4.21, in which an LED is connected through a switch to a fixed DC supply voltage, V_{on}.

The time it takes for an entire switching cycle, τ_{swc}, consists of the time that the LED is on, τ_{on}, and off, τ_{off}:

$$\tau_{swc} \equiv \tau_{on} + \tau_{off}. \tag{4.1}$$

4-21

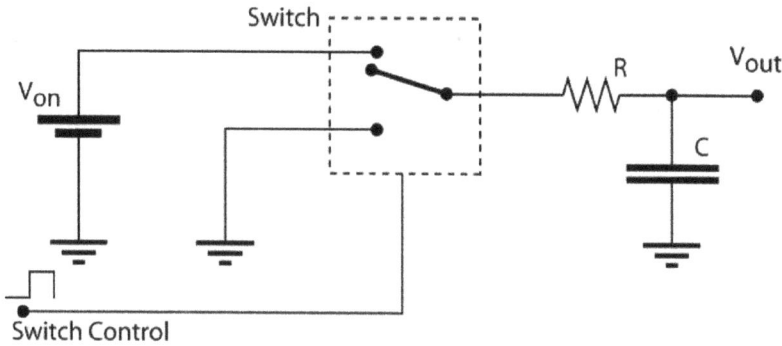

Figure 4.22. PWM circuit with an external (electronic) switch control to turn the switch on or off.

The switching (cycle) rate or speed, f_{swc}, is defined as

$$f_{swc} = 1/\tau_{swc}. \tag{4.2}$$

At first, we will operate the switch by hand. As long as τ_{on} and τ_{off} are on the order of seconds, our eyes will observe the LED as either fully on or fully off.

Next, consider what happens when we connect the switch control to a function generator. Assume for now that the time interval that the light remains on, τ_{on}, is equivalent to the time it is off, τ_{off}.

At a low switching rate, at a few hertz, our eyes observe the familiar full on/off behavior. As the switching speed is increased, we begin to notice the blinking LED. As the switching speed is further increased, we reach a point where the response of our eyes, specifically the retina retention, is too slow to register the LED's distinct on and off states. Instead, at this frequency and beyond it, our eyes will act like an 'averager' or low-pass filter. We perceive an average intensity that is somewhere between the LED's full brightness and it being completely turned off. In the current case, with $\tau_{on} = \tau_{off}$, we probably perceive the LED at half its maximum brightness.

How does the ratio of τ_{on} and τ_{off} affect the 'average' brightness perceived? Assume that the switching rate, f_{swc}, is kept constant and remains at a frequency exceeding our eyes' response time. As mentioned previously, at this switching rate, the perceived intensity will be an average based on how long the LED is on or off during each switching cycle. For example, if $\tau_{on} \gg \tau_{off}$, the LED is on most of the time and a bright LED will be seen; for $\tau_{on} \ll \tau_{off}$, the LED is mostly off and a dimly lit LED will be observed. It follows that by controlling the ratio of τ_{on} and τ_{off} we can obtain any desired level of intensity.

Let us determine a mathematical relationship between τ_{on} and τ_{off} and the 'perceived' average intensity. An analogous model of the eye and its response can be represented by the circuit shown in figure 4.22. In this model, V_{on} would be analogous to the maximum light intensity; the RC low-pass filter acts like an averager, representing our eyes; finally, $\langle V_{out} \rangle$ is the average output voltage and it corresponds to the perceived intensity. Therefore, we would like to calculate $\langle V_{out}(RC, \tau_{on}\tau_{off}) \rangle$.

As will be calculated in this section, as long as RC is much larger than the time for one switching cycle, that is, RC $\gg \tau_{swc}$, $\langle V_{out} \rangle$ is

$$\langle V_{out} \rangle = \tau_{on} V_{on}/(\tau_{on} + \tau_{off}) \tag{4.3a}$$

$$= \tau_{on} V_{on}/\tau_{swc}. \tag{4.3b}$$

Before we proceed to implementing the PWM circuit in Verilog, let us consider the implications of the results above.

First, these results show that with the aid of a timing signal controlling a switch, a fixed (DC) voltage can be converted into a voltage that is less than its maximum input voltage, V_{on}. In other words, for RC $\gg \tau_{swc}$, the circuit acts very much like a voltage divider.

Second, if the system to which the PWM is applied responds much more slowly than the switching speed, then the low-pass filter that provides the averaging is no longer required. For example, when using the LED and our eyes, the eyes will provide the low-pass filtering. This is true for many other cases. For example, in 'power' applications where the PWM technique is used to control motors, solenoids, actuators or the speakers for our audio player, as long as the response time of the device being powered is much slower than τ_{swc} the low-pass filter is no longer required. In all these cases the slow response of the system that is being controlled provides the filtering indirectly.

Combining these two observations leads to a very important implication: since the PWM circuit acts like a voltage divider and if the low-pass filter is part of the 'load' then all the power will be absorbed by the load and none will be wasted by the regulating circuitry. In other words, a PWM circuit can be far more efficient in regulating power than a regular two resistor voltage divider where a lot of power is being wasted in the regulating resistors (see review exercise 4.5.5., question 2.). This is one reason for the popularity of PWM in many low power applications such as portable electronics devices.

4.5.3 Simple PWM technique

A critical element in the PWM technique is the switch control algorithm or module; it controls the time interval that the switch stays on, τ_{on}, or off, τ_{off} (see the diagram in figure 4.23). This module has two inputs. An external clock input, f_{clock}, provides a trigger signal for the algorithm and also acts as a reference clock signal. The 'x_in' input controls the number of reference clock cycles that the switch stays on or off. In other words 'x_in' will be directly proportional to τ_{on}, and hence $\langle V_{out} \rangle$. Its output, 'PWM_out', is the PWM signal. It can be used directly to control low power applications such as dimming an LED, but more typically, this signal is fed into some sort of physical switch, for example a transistor or relay, to provide power amplification, as shown in figure 4.23.

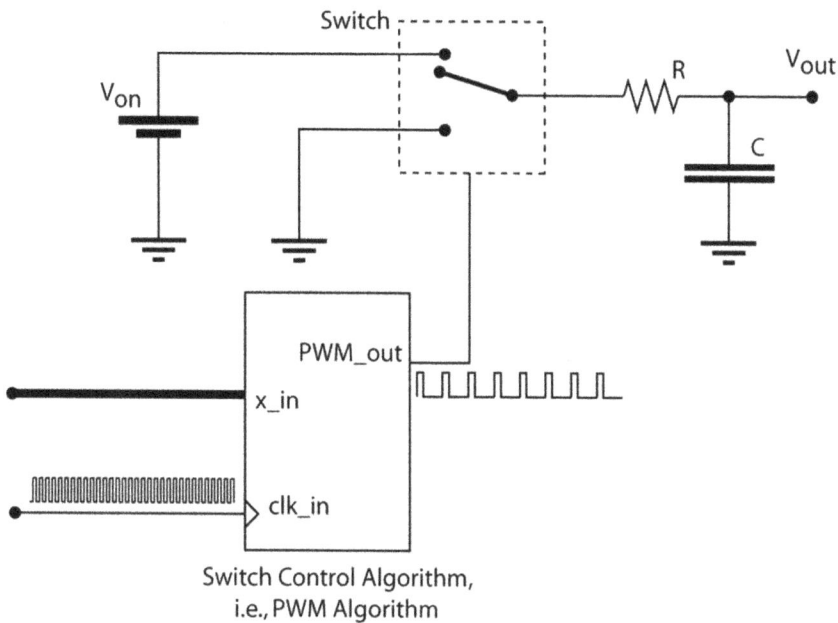

Figure 4.23. The complete PWM circuit. The actual switching element can be a physical switch, such as an electrical relay. However, transistors are typically used for switching, most commonly in the configuration of a silicon controlled rectifier (SCR). The low-pass filter may be omitted if the system to which the PWM signal is applied responds much more slowly than one complete switching cycle.

Now let us look at the switch control algorithm. After this lengthy discussion, it may come as a pleasant surprise to discover just how straightforward it is. It can be expressed using the following pseudo-code

```
if ( counter < x_in )

                   PWM_out <= 1;

         else

                   PWM_out <= 0;

counter <= counter+1;
```

While the pseudo-code shown above does not have the proper Verilog syntax, it is close to it so you should have no problem implementing it in this exercise.

As indicated in the pseudo-code above, the main components of the PWM technique are a counter and a digital comparator. Whenever the continuous count value exceeds the input, 'x_in', the module outputs a low 'PWM_out' signal, otherwise it is high. In other words, its output, 'PWM_out', acts like the switch output from the previous discussion.

The input, 'x_in', is a user selectable value. It directly controls how long the output remains high or low, that is, how long the 'switch' stays 'on' and 'off'. Thus,

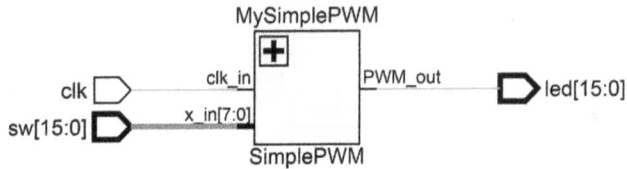

Figure 4.24. Elaborated schematic of the simple PWM module.

it determines the length of τ_{on} and τ_{off}. The range of 'x_in' must never exceed the largest count value of the counter. Therefore, its vector size is identical to that used for the counter.

Note that this is a free running counter. Once it reaches its maximum value, it starts over again at zero. Since the counter is incremented at a frequency, f_{clock}, the duration of an entire switching cycle for an n-bit counter is

$$\tau_{swc} = 2^n/f_{clock} \, . \tag{4.4}$$

From this it follows that the switch remains 'on' for the following time interval:

$$\tau_{on} = x/f_{clock} \, . \tag{4.5}$$

If the output from the module were connected to a low-pass filter with a large RC time constant (see figure 4.22), then we would find that $\langle V_{out} \rangle$ is

$$\langle V_{out} \rangle = \text{x\_in} \, V_{on}/2^n. \tag{4.6}$$

In other words, the averaged output voltage is directly proportional to 'x_in'. Since 'x_in' is a digital value and $\langle V_{out} \rangle$ is an analog value, we see that the PWM circuit acts like a *digital-to-analog converter*, that is, a D2A.

4.5.4 Exercise 1

Implement the simple PWM algorithm shown above using an 8-bit counter. First write a standalone module; its schematic is shown above in figure 4.24. Instantiate your PWM module in a new top module using switches SW[7] through SW[0] for 'x_in' and assign 'PWM_out' to led[0]. Additionally, add another port to your top module, [7:0]JB, and assign 'PWM_out' to JB[0], i.e. pin 1 of connector JB. This way, you will be able to observe the PWM output signal on LED0 and the scope.

Add a copy of the master constraint file and uncomment the appropriate lines. Generate the bit-file and program your board. Connect a BNC1 Pmod to the (upper row) of the Pmod JB connectors on your BASYS3 board. (See figures 4.15 through 4.17 in the previous section.) Check that its blue jumper is in the A position and connect the Pmod's output to the scope.

Can you control the intensity of LED0 using switches SW7 through SW0? Observe the corresponding signal on the scope. Does it make sense?

4.5.5 Review exercises

1. For RC $\gg \tau_{swc}$, calculate $\langle V_{out}(\text{RC}, \tau_{on}, \tau_{off}) \rangle$ using the circuit shown in figure 4.22. You may use either of the two methods outlined below:

Hint 1. Calculate the average by using the general form: $\langle f(t) \rangle = \frac{1}{t_0} \int_0^{t_0} f(t) \mathrm{d}t$ where $f(t)$ corresponds to the voltage fed *into* the low-pass filter during one entire cycle, that is $t_0 = \tau_{\mathrm{on}} + \tau_{\mathrm{off}}$. Find $f(t)$ for $0 \leqslant t < \tau_{\mathrm{on}}$ and for $\tau_{\mathrm{on}} \leqslant t \leqslant \tau_{\mathrm{on}} + \tau_{\mathrm{off}}$. Since the low-pass filter is an averaging circuit, the result obtained by this method is identical to the one obtained using an ideal low-pass filter.

Hint 2. Find a differential equation for the change in voltage, $\mathrm{d}V$, across the capacitor during $\mathrm{d}t = \tau_{\mathrm{on}}$ and $\mathrm{d}t = \tau_{\mathrm{off}}$. Note that V_{in} during τ_{on} is V_{on} and it is 0 during τ_{off}. Finally, for the output to remain around a fixed (steady state) value $\langle V_{\mathrm{out}} \rangle$, $\mathrm{d}V$ during the on time must be equal to $-\mathrm{d}V$ during the off time.

2. To conserve energy, a person decides to dim the lights in the house by adding a regulating resistor in series with a lamp. For now, assume that the resistance of the regulating resistor is five times the resistance of the lamp. How much power overall was saved by the entire circuit? How much more power is absorbed by the regulating resistor then by the lamp? Why would a PWM circuit be a better choice?

3. Two fundamental characteristics of any D2A (or A2D) are its resolution (also called 'sensitivity') and its conversion speed. For a D2A, like the one used here, the resolution or sensitivity is defined as the smallest voltage change in its output when we change its input. Calculate this value for your 8-bit PWM module. (*Hint*: check Equation (4.6) and assume $V_{\mathrm{on}} = 3.3$ V.)

4. Calculate the resolution or sensitivity if you were to change the counter in your algorithm to a 16- or 64-bit counter.

5. The conversion speed is the time required for a D2A (or an A2D) to adjust its output. While our simple PWM algorithm is updated at each system clock cycle, f_{clock}, we have to wait (at least) one complete τ_{swc} interval to average the signal. Hence, we assume τ_{swc} to be its conversion speed. For the BASYS3 board, with $f_{\mathrm{clock}} = 100$ MHz, what is the algorithm's conversion speed if it uses an 8-bit counter?

6. What would the simple PWM algorithm's conversion speed be for the BASYS3 board if it were to use a 16- or a 64-bit counter? (Please convert the latter result into years.)

7. Nyquist's theorem states that to truthfully recover the frequency of a signal it must be sampled (or played back) at (at least) twice its original frequency. From, (4.5) and (4.6), what are the maximum frequencies that can be truthfully recovered for the simple PWM algorithm with an 8-, 16- and 64-bit counter?

8. Create the Verilog code for exercise 1 and show a screenshot from the scope when x_in = 8, 128.

4.6 An audio player with PWM

4.6.1 Introduction

In this exercise you will finally assemble all the components for the simple audio player. (A complete elaborated schematic diagram is shown below.) It works by reading one byte of digital audio data from a (preprogramed) flash memory card at a frequency of 44.1 kHz. This information is then converted to an analog signal using a PWM algorithm. Finally, the analog signal is sent to an audio amplifier and speaker connected to the BASYS3 board.

4.6.2 Detailed description

Here is a detailed description of each module and its functionality; please compare these with the schematic shown in figure 4.25.

1. 'AudioClock'. The purpose of this module is to reduce the system clock from 100 MHz to an audio frequency of 44.1 kHz. It uses a counter and the appropriate logic conditions to send out a 10 ns pulse every 1/44.1 kHz. The digital audio data will be read from the flash memory card at this audio frequency and it is subsequently sent to the PWM A2D converter. You have already developed and tested this module in section 4.4.

2. 'AddressCounter'. This module generates the address for the one byte of digital audio data to be read from the flash memory. It is a 24-bit counter which is incremented each time its 'clk_in' input goes high, i.e. the output of the 'AudioClock' module. The counter is configured to increment until it reaches a value of 24'h1F_FFFF (this notation means: 24 bits with hex value: 1F_FFFF) at which point the counter resets itself to 0. You will have to create this module.

3. 'ReadOneByteV3'. This module reads one byte of digital audio data from the flash memory at the address specified at its 'address' input. The byte is read each time its 'enable' input goes high. The 'ReadOneByteV3' module has already been written and will be provided to you. It can be found in the file 'FlashMemModulesV7.v' downloaded from: https://sites.google.com/a/umn.edu/mxp-fpga/home/vivado-notes/phys4051-course-related-materials.

4. 'SimplePWM'. This module converts the 8-bit digital audio data read from the flash memory to an analog signal using pulse width modulation. The analog signal is then amplified and played on a speaker connected to the BASYS3 board. You have already developed and tested this module in section 4.5.

As you can see, except for writing the top module and instantiating the various modules, the only new module to be developed is #2, the 24-bit 'AddressCounter'. Let us start by developing it in a new project.

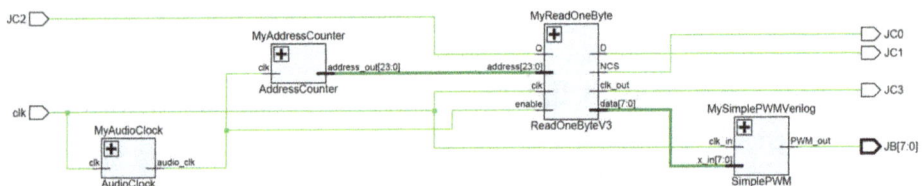

Figure 4.25. Elaborated schematic design of the complete audio player (top) module with all its instantiated modules.

4.6.3 'AddressCounter' module

This module works like the decade counter discussed in section 4.4.2. The difference being that this counter uses a 24-bit register and that it is reset to 0 each time it reaches 24'h1F_FFFF. However, unlike the decade counter, this module does not require a 'c_out' output.

4.6.4 Create the top module and instantiate the modules

Copy all the required files into your current project. The 'ReadOneByteV3' is found in the folder https://sites.google.com/a/umn.edu/mxp-fpga/home/vivado-notes/phys4051-course-related-materials in file 'FlashMemModulesV7.v'. Copy your Verilog code for the audio clock and PWM from sections 4.4 and 4.5 into the current project.

Create the header for your new top module; a listing of all its external ports and their direction can be found in the schematic diagram in figure 4.25.

Start by instantiating the 'AudioClock' and then the 'AddressCounter'. The output of the 'AddressCounter' is an internal *24-bit wire vector*; make sure you declare this vector explicitly in your code!

Next instantiate the 'ReadOneByteV3' module. Again be careful to explicitly declare the 8-bit internal wire for 'data'.

Finally instantiate your 'SimplePWM' module and assign 'PWM_out' to JB[0].

Synthesize your project and check that it contains no syntax errors.

4.6.5 Physical connections and the constraint file

From the photograph in figure 4.26 you can see that the Pmod Digilent PModSF 2 MB flash memory module has been connected to the JC connectors, at the lower

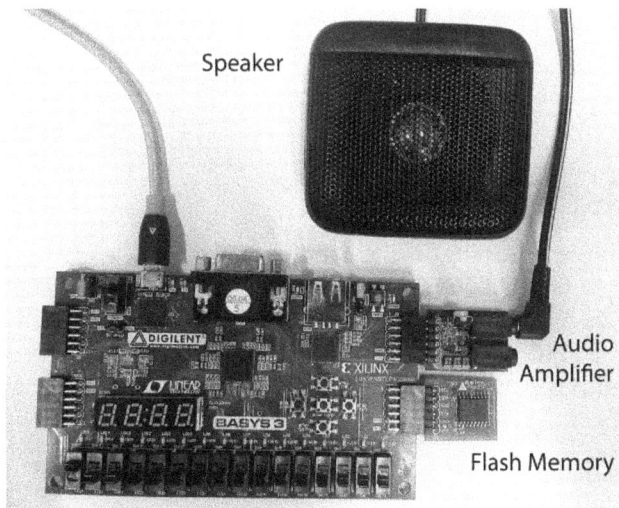

Figure 4.26. BASYS3 board with the audio amplifier and its speaker and the flash memory. (You could substitute a pair of headphones for the speaker).

right-hand side of the BASYS3 board; the Digilent PmodAMP1 speaker/headphone amplifier module to boost the audio signal strength has been connected to the JB connectors at the upper left corner of the board.

Both modules must be plugged into the *upper* row of the Pmod connectors, i.e. utilizing Pins 1 through 6, with their component side (as shown in figure 4.26) facing up.

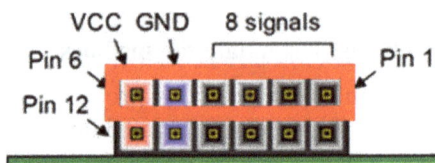

Figure 4.27. Be sure to connect the audio amplifier and flash memory to the upper row of pins of the Pmod connector JB.

Copy the BASYS3 master constraint file into your project and edit it. From the port declarations you can see that you need to uncomment the statements for the system clock, 'clk', and the JB[7:0] connectors.

At this point, the only remaining pin mappings are the ones for the Pmod JC connectors to the flash memory card. However, since some of these signals are inputs (JC[2]) and others (JC[0], JC[1] and JC[3]) are outputs we cannot declare JC a vector. (Remember, by definition, a vector is always unidirectional, i.e. all its components are either inputs or output but never a mix of both.) Instead, we must modify each of their entries in the constraint file to indicate that they represent single-bit ports.

After uncommenting the statements for the first four JC connectors, i.e. JC[0] through JC[3], remove the square and curly brackets so that JC[0] becomes JC0 and JC[1] JC1, etc. The edited version of this part of the constraint file should look like:

```
set_property PACKAGE_PIN K17 [get_ports JC0]
    set_property IOSTANDARD LVCMOS33 [get_ports JC0]
#Sch name = JC2
set_property PACKAGE_PIN M18 [get_ports JC1]
    set_property IOSTANDARD LVCMOS33 [get_ports JC1]
#Sch name = JC3
set_property PACKAGE_PIN N17 [get_ports JC2]
    set_property IOSTANDARD LVCMOS33 [get_ports JC2]
#Sch name = JC4
set_property PACKAGE_PIN P18 [get_ports JC3]
    set_property IOSTANDARD LVCMOS33 [get_ports JC3]
```

Generate the bit-file and plug in the 8-bit flash memory and the audio amplifier board as shown in figure 4.26. The speaker works only when it is plugged into the input closest to the round trimpot. This trimpot on the PmodAMP1 module can be used to adjust the audio volume; turning it all the way counter-clockwise produces the maximum output power. Try the different 8-bit audio recordings and enjoy.

4.6.6 Additional details

The following information has been provided for your information in case you want to learn more about the details of the projects.

Table 4.2. Description of the flash memory's inputs and outputs.

Signal name	Direction	Purpose	Six-pin edge connector JX pin assignment
Q	Input	Read data from memory line	3
D	Output	Send data to memory line	2
NCS	Output	Chip select line (negative logic)	1
CLK_OUT	Output	Clock line	4

Audio data. The audio data were created from 8- and 16-bit (mono) WAV files sampled at 44.1 kHz. WAV files are easy to work with for data manipulation because they have not been compressed; the audio information is stored in a sequential binary file which can be read by a C-program and represents the digitized audio signal amplitude as a function of time. Unfortunately, since the information has not been compressed, it uses a lot of memory. This is the reason why we are only able to play such short audio pieces. (Of course, it would far more rewarding if we were able to use a MP3 compression/decompression algorithm for our data. Unfortunately, these codes are copyrighted and not easily obtainable.)

Flash memory. After the binary data have been read by the C-program they are offset to the middle of their dynamic range (to compensate for signed values) and then they are stored (via an RS232 interface and the BASYS board) in the flash memory.

The flash memory can contain up to 2 MB of data and it communicates with the BASYS board through a serial peripheral interface (SPI) using four pins listed in table 4.2.

Although it does not matter which of the four six-pin edge connectors (JA through JD) you will be using for the flash memory cards, it is crucial that once you have selected one, you assign the signals above to the appropriated pins. For this exercise, we use connector JC.

'ReadOneByteV3' module. This module will read one byte of data from the flash memory at the address specified at its input. The module uses the ports listed in table 4.3.

Audio hardware. The Digilent PmodAMP1 speaker/headphone amplifier amplifies low power audio signals to drive either a stereo headphone or a monophonic speaker. The speaker is driven from the input closest to the round trimpot. The PmodAMP1 module is connected to the BASYS board through PMod connector JB. Use pin 1 for the PWM signal. The round trimpot on the PmodAMP1 module can be used to adjust the audio volume; turning it all the way counter-clockwise produces the maximum output power.

Table 4.3. Description of the 'ReadOneByteV3' module ports.

Direction	Name	Size (bits)	Description
Input	clk_in	1	Serial interface clock for transmitting data; use the 25 MHz system clock.
Input	enable	1	Hardware trigger: at the positive edge, the new data are retrieved and subsequently stored in registers 'data'.
Input	address	24	Address where data are retrieved from in the flash memory.
Input	Q	1	Data read from memory are transmitted through this line.
Output	NCS	1	(Not) chip-select line.
Output	D	1	Data sent to the memory are transmitted through this line.
Output	clk_out	1	Clock signal from memory; required when receiving data.
Output	Data	8 or 16	Depending on version, either 8- or 16-bit register holding the data value read from memory.

Additional information regarding the flash memory and audio amplifier modules can be found here:

Memory: http://www.digilentinc.com/Data/Products/PMOD-SF/PMod_SF_%20rm.pdf

Audio: http://www.digilentinc.com/Data/Products/PMOD-AMP1/PmodAMP1_rm_RevB.pdf

4.6.7 Review exercises

1. Describe the major components (and their function) of the audio player.
2. Create a well-documented version of your Verilog code for the 8-bit audio player.
3. As you may recall from the address counter, the flash memory can store 24'h1F_FFFF + 1 bytes of audio data. How long will it take your audio player to play them once?

Chapter 5

Counters

Additional reading

Read the following pages/section from Horowitz P and Hill W 2015 *The Art of Electronics* 3rd edn (New York: Cambridge University Press) for the sections of this book indicated:

- pp 737–40 for section 5.1.1.
- pp 736–7 for section 5.2.
- skim through section 15.1 (p 1053) for section 5.4.

5.1 Introduction

In this last chapter on Verilog, you will build a complete function/event counter and a period counter that you will then interface with the lab computer. While a large part of the exercise will expose you to concepts related to frequency and period counters, it will also give you a chance to explore new tools and capabilities for designing and working with FPGAs. Hopefully, you will obtain an appreciation of how powerful and limitless its capabilities are.

5.1.1 Synchronous versus asynchronous sequential logic

An event or frequency counter (see figure 5.1), like the one you will be building in this chapter, records the number of events during some reference time interval, τ_{ref}. This reference time interval is obtained by counting a fixed number of clock cycles, n, of a stable clock with frequency f_{clk}:

$$\tau_{\text{ref}} = n/f_{\text{clk}}. \tag{5.1}$$

The timing diagram below shows such a counter which uses a 100 MHz system clock, like the one on the BASYS3 boards, as the reference clock, f_{clk}; τ_{ref} consist of

5-1

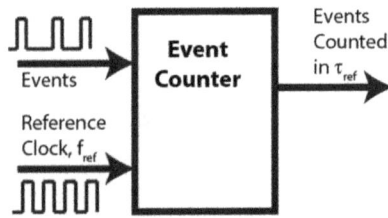

Figure 5.1. Event counter schematic.

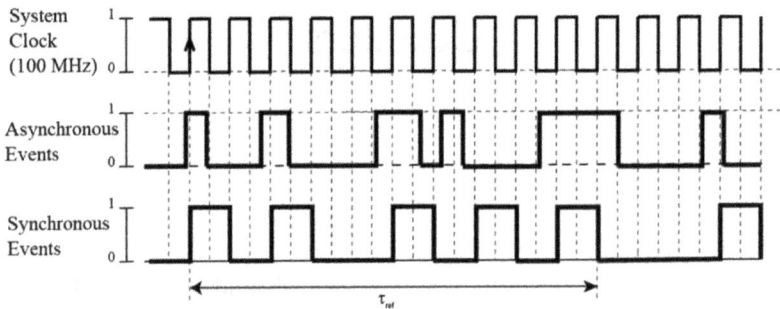

Figure 5.2. Timing diagram for an event counter with the reference clock (system clock) and asynchronous events and the same events synchronized.

ten system clock cycles and, therefore, would count events during a 100 μs time interval. In the example in figure 5.2, this corresponds to five events or an event rate of 50 kHz.

Events occur either 'in-sync' or 'out-of-sync' with respect to the reference clock. If they are in-sync, i.e. synchronous, then all their active edges align with those of the reference clock. However, as is far more common when we count events, most are asynchronous or out-of-sync with the reference clock. For example, in particle counting, the events occur at random time intervals (and sometimes even duration) as shown in the second trace in figure 5.2. Even for periodic signals, for example if we measure the frequency of a square wave, it is highly unlikely that all its active edges line up with those of the reference clock.

The Verilog implementation for such an event counter depends to a large extent on the synchronicity of the input clock signal. Although Verilog allows the use of asynchronous clock signals by using multiple clock signals in its always-block, it is best to avoid them whenever possible. The compiler will complain because they can cause race-conditions and glitches that are very difficult to diagnose.

For illustration only, shown below is a simple extension of the previously shown D-type module with a reset input. While the reset in the example below does not have to be synchronous, its effect on the flip-flop will be synchronous since the always-block updates only on the positive edge of the clock signal.

```
module MyFDR(  //synchronous reset
    input D_in,
    input clk,
    input reset,
    output reg Q_out);

always@( posedge clk) begin
    if( reset )
            Q_out <= 0;
    else
            Q_out <= D_in;
    end
endmodule
```

However, the second example below uses an asynchronous reset input. The always-block is now controlled by *two* clock signals, namely 'clk' and 'reset' and its output may change at the moment either one of these inputs is asserted. Furthermore, from the conditional logic, 'reset' dominates 'clk'; once it is high it essentially overrides 'clk'. Finally, a large reason why the Verilog compiler does not like these asynchronous clocks is because it has no way of predicting what the outcome of the always-block is when the 'clk' and 'reset' events happen simultaneously or very close together.

```
module MyFF(  //asynchronous reset
    input D_in,
    input clk,
    input reset,
    output reg Q_out);

always@( posedge clk or posedge reset) begin
    if( reset )
            Q_out <= 0;
    else
            Q_out <= D_in;
    end

endmodule
```

In short, avoid using multiple clock signals, i.e. asynchronous clocks, to control the always-block. In the next section you will learn how to convert asynchronous clock signals into synchronous ones using a one-shot/synchronizer module.

5.2 One-shot or monostable multivibrator and synchronizer and Vivado's behavior simulation tool

A 'one-shot', or more technically a 'monostable multivibrator', circuit produces an output pulse of a fixed duration when its input has been triggered. In other words, it behaves like an old fashioned flush toilet which flushes for a fixed time interval once it has been 'activated'.

One-shots come in two flavors: retriggerable and non-retriggerable. To use our plumbing analogy, the distinction involves how the system handles another flushing request while it is still busy flushing. In the first case, the retriggerable one-shot, the system would keep on flushing, while in the second case, the non-retriggerable case, the system ignores any new requests while it is busy.

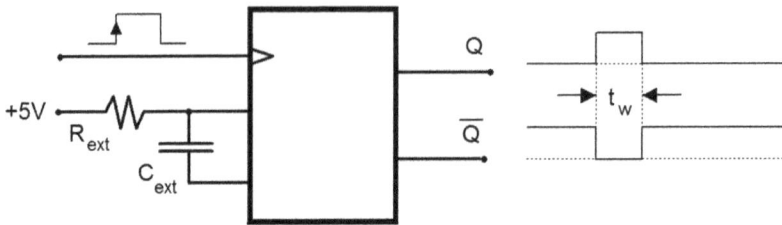

Figure 5.3. (Simplified) block diagram of a monostable multivibrator.

'Converting' input pulses to output pulses of identical duration has many applications. For example, if the signals are of extremely short duration, as is typical in particle detectors, 'stretching' them to a fixed length ensures that they can be detected by the electronics. On the other extreme, very long pulses, such as those that might be created from a human activated push button, may tie up a circuit for an indefinite duration; reducing the pulse to a fixed length allows the system to process the human input a fixed time after it has been received. In summary, it is simply much easier to design a system where all signals are of the same duration and where we can control that duration by, for example, using a one-shot.

One-shots exist as ASSP chips, such as the popular 74123 IC shown in figure 5.3. It contains an internal clock and the duration of the output pulse, t_w, is controlled by an external resistor, R_{ext}, and capacitor, C_{ext}. The fixed width output pulse is generated at the rising edge of its clock input.

5.2.1 One-shot with synchronizer Verilog module

Shown below is a digital implementation of the one-shot circuit of figure 5.3. You will now use Vivado's built-in simulator to create a timing diagram of this one-shot/synchronizer module. This Vivado tool is extremely useful for testing, debugging and analyzing code. You will use it to figure out what the code shown below does and how it works.

```verilog
module OneShotV4(
    input clk,
    input asynctrigger_in,
    output reg pulse_out = 0
    );

    reg trig_set = 0;

    //asynchronous flip-flop
    always@(posedge asynctrigger_in or posedge pulse_out)
            if( pulse_out)
                    trig_set <= 0;
            else
                    trig_set <=1;

    always@( posedge clk)
            if( trig_set)
                    pulse_out <= 1;
            else
                    pulse_out <=0;

endmodule
```

Start a new project and add the file named 'OneShotSyncV4.v' from the folder https://sites.google.com/a/umn.edu/mxp-fpga/home/vivado-notes/phys4051-course-related-materials or enter the code shown above.

Start the timing simulation by selecting: 'Simulation/Run Simulation/Run Behavioral Simulation' (see below).

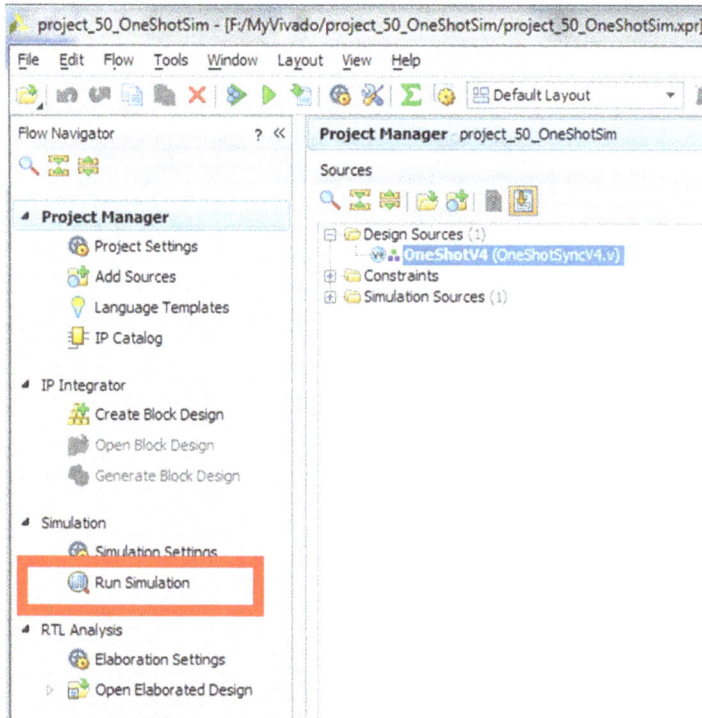

After a few seconds, or more, the timing diagram window, or 'wave window', shown below (box 1) opens.

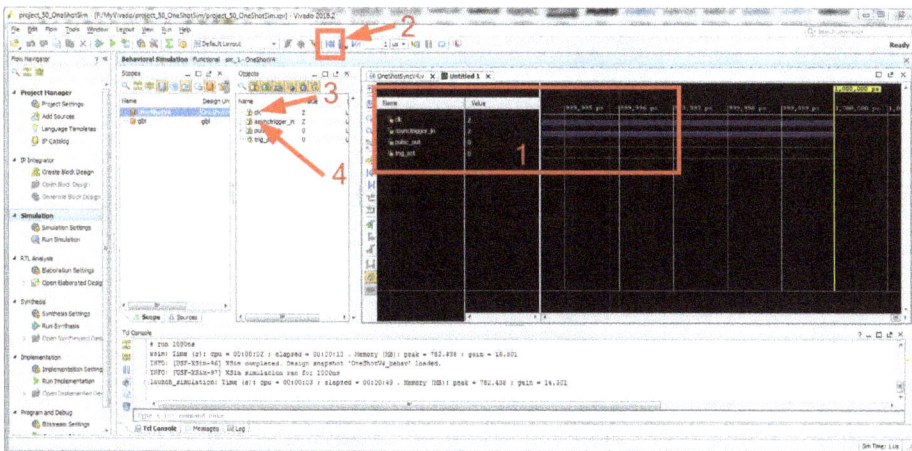

You will notice that all its input signals are permanently frozen in a 'z' state, indicating that the input states were not specified. ('z' stands for 'high-z state', i.e. infinite input resistance, or open circuit.) Of course, without knowing the state of the input signals, the simulator is not able to determine the outputs.

So let us start the simulation over by pressing the 'Restart' button, (2). (This is optional; however, it is nice to have the simulation starting at $t = 0$.)

Next, specify your clock signal. Right-click on 'clk' (3) and select 'Force Clock'. The window below opens.

We want the clock to start low (labeled: 'Leading edge value') and then go high ('Trailing edge value'). After entering the appropriate values, set the period to our BASYS3 system clock, 10 ns, and hit 'OK'.

Finally, add a forced clock signal to the 'asynctrigger_in' input (4 in the screenshot previous to the above). Since we want this input to be asynchronous with the system clock use 'weird' values for its period and add an arbitrary 'Starting (after) time offset'. See below and hit 'OK' when you are satisfied.

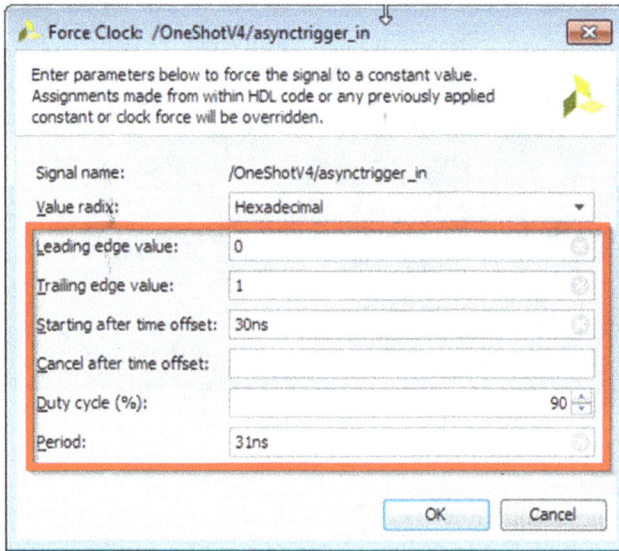

Back in Vivado's IDE, change the 'Run for 10 µs' window to 1 µs and then press the run button on its left (1 in screenshot below). Since the wave window always displays the last few nanoseconds of the simulation, press the 'Zoom Fit' button (2) to see the entire trace. The timing diagram should look similar to the one shown below.

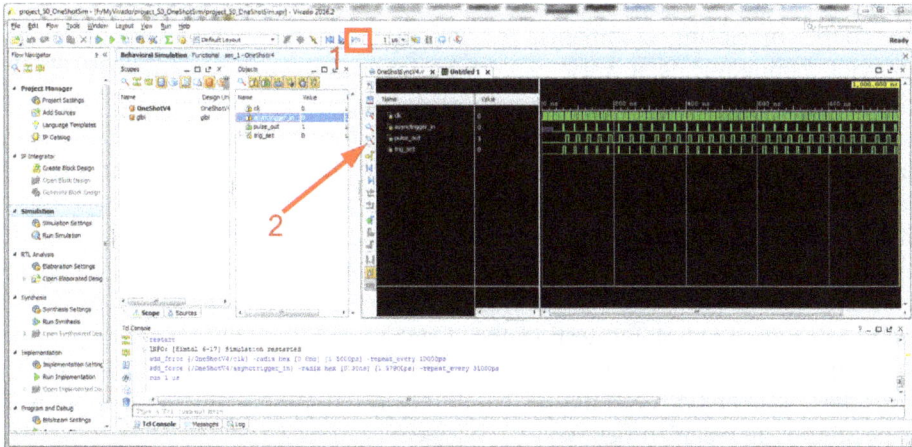

Use the combination of the 'Zoom In' tool (1 below) and the slider (2) to obtain a better image of the trace. Not only does the trace show your input and output signals, it also lists the state of your internal signals, in this case 'trig_set'.

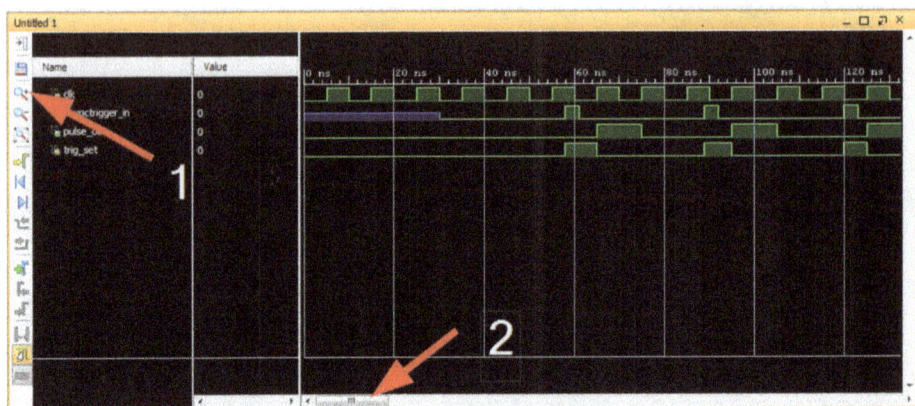

From this you should now be able to answer some questions about the code provided, such as, how long are the output pulses (with respect to the input clock) and are they in sync with the input clock?

Finally, determine if the 'OneShotV4' module acts like a retriggerable or non-retriggerable one-shot. Modify the values for the 'asynctrigger_in' forced clock to obtain multiple trigger signals in the single system clock cycle and run the simulation for another 1 μs. (*Warning*: pressing the 'Restart' button resets all the input stimuli signals back to high-z!) Unless you also want to re-enter the values for the forced clock 'clk' input, enter only the values for the 'asynctrigger_in' forced clock and then press the 'Run for 1 μs' button. The new simulation will be appended to the previous one.

A couple of closing comments. As you may have already noticed, Vivado's 'Behavior Simulator' is a very powerful tool but it is also extremely clunky, especially when you use it the way we just did. If you were debugging your code you would have to rerun the 'Behavior Simulator' after each change and re-enter all the input stimuli again. That quickly becomes very cumbersome and annoying. As you may have guessed, there is a far better method and it involves writing scripts or test benches which specify the input stimuli and how they change during the simulation. The commands for these stimuli are a part of the Verilog language. However, writing such test scripts is outside the scope of this book. If you are curious, a sample script for testing the 'OneShotV4' module has been provided below for illustration only. If you were to add it to your project as a simulation source file, the 'Behavior Simulator' would always automatically start up with the timing stimuli as specified below.

```
module MyTb;
    reg clk;
    reg asynctrigger_in;
    wire pulse_out;

    //Instantiates the Unit Under Test (UUT)
    OneShotV4 uut(.clk(clk), .asynctrigger_in(asynctrigger_in), .pulse_out(pulse_out));

    initial begin   //initial conditions; to be executed only once
        clk = 0;
        asynctrigger_in  = 0;

        #13   asynctrigger_in  = 1;  //means: wait 13 nsec then set this signal high
        #3    asynctrigger_in  = 0;  //means: after 3 nsec set this signal low, etc.

        #2    asynctrigger_in  = 1;
        #2    asynctrigger_in  = 0;
    end

    //continuous clock signal
    parameter CLOCKPERIOD = 10;
    always begin
        clk = 1'b0;
        #(CLOCKPERIOD/2) clk = 1'b1;
        #(CLOCKPERIOD/2);
    end

endmodule
```

Finally, as you may have noticed, the 'OneShotV4' module contains two always-blocks and one of them is an asynchronous flip-flop because it has two clock signal inputs. We previously emphasized employing them with caution; however, in this case it is difficult to get a digital circuit working without their aid.

5.2.2 Review exercises

1. Describe the 'OneShotV4' module. What is its intended purpose and how does it work? Use timing diagrams from the 'Behavior Simulator' to illustrate your statements; specifically show if or how its output is synchronized with the system clock (or not) and if the one-shot is retriggerable or not.

2. When two one-shots are connected as shown in figure 5.4, they become an oscillator circuit. Try to understand why. Assume that both one-shots are triggered on the positive edge of their clock input. If the first one shot creates a positive output pulse (at Q_1) of a 1 ms duration and the second one-shot a 9 ms positive pulse (at Q_2), then at what frequency will this circuit oscillate?

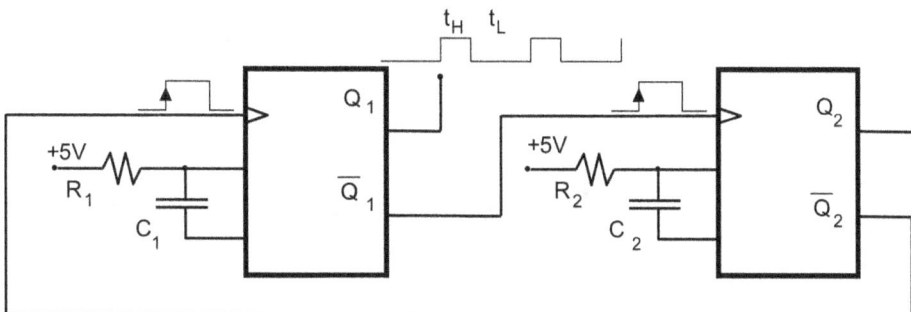

Figure 5.4. Two one-shot circuits configured as an oscillator.

5.3 State machine frequency counter

Our approach for creating the various projects has been modular: we create modules, test them individually and then instantiate them all together in a top module. An example of this is shown in figure 5.5, the schematic of the frequency counter which you will implement in this section. It consists of only three modules of which two are already familiar to you.

As the complexity of the tasks increases, it can become difficult to create compact modules which can easily be tested and debugged. In such situations, programmers often describe and implement the task by resorting to *finite state machines* (FSMs). They are well understood and clarify the sequential designs by using centralized states. We will follow this approach and implement the event counting module, named 'SyncEventCounterV1' in the schematic above, using such a state machine.

5.3.1 Finite state machines (FSMs)

Before we discuss the FSM diagram in figure 5.7, we will cover some basic concepts. (We will use a Moore type FSM, although most of the concepts would equally apply to a Mealy type FSM. For the sake of brevity, we will not discuss the difference between these two types.)

Figure 5.6 highlights the key components of an FSM: the FSM is driven by a clock signal and its states may only change at the active clock edge. Also, for a

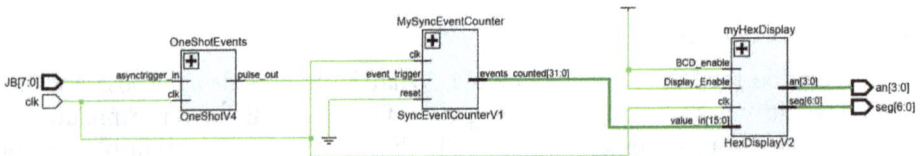

Figure 5.5. Complete event counter schematic. Use JB[1] for the 'asynctrigger_in' signal.

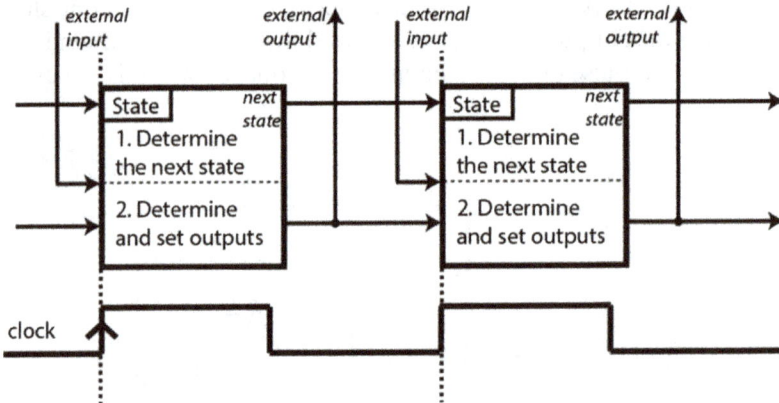

Figure 5.6. Time evolution of a state machine for two clock cycles; the dotted vertical lines indicate the start of a new state.

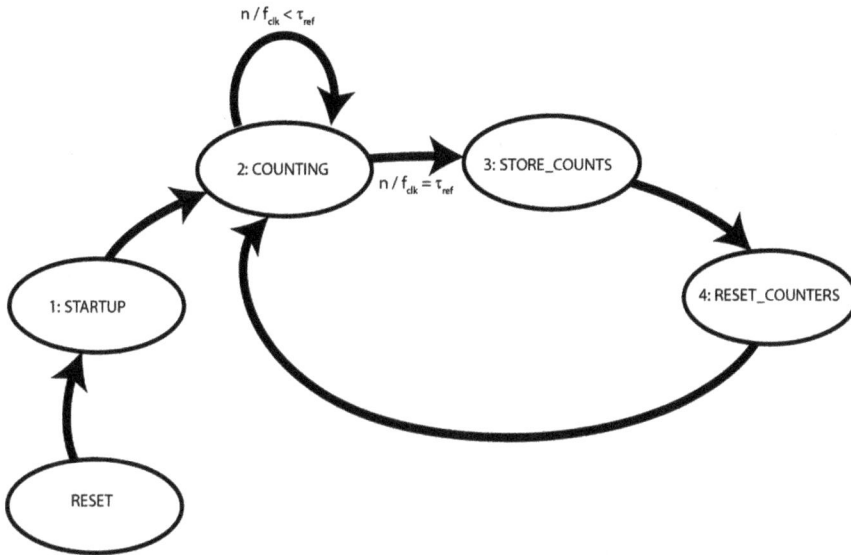

Figure 5.7. State machine diagram for our event counter.

Moore type FSM, on the active edge, all its input values are latched. These signals have been set in the previous state or they may originate from external inputs.

While the FSM is in a particular state, it completes two tasks:

1. The FSM determines the next state. This decision is based on the combinational logic of its latched input signals.
2. The FSM sets its various output registers and/or counters. Again, this decision is based on the combinational logic of its latched input signals.

An important point about selecting the next state: as can be seen from figure 5.7, the FSM may decide to remain in its current state, for example in the COUNTING state. In that case, it will remain in the COUNTING state for one more clock cycle and re-evaluate its next state based on the new input signals. This process may repeat itself until the input conditions are finally favorable to move on to a new state. In the example below, the FSM will break out from its COUNTING state once its reference time period τ_{ref} has elapsed.

5.3.2 Event counter FSM

Let us now examine the state diagram for our event counter, shown in figure 5.7, in more detail. Assume that the events we are counting have all been 'cleaned-up' by the one-shot/synchronizer circuit from the previous section and that they all last exactly one system clock cycle. Furthermore, assume that the FSM is driven by the BASYS3 100 MHz system clock, f_{clk}, making each state active for exactly 10 ns.

Table 5.1. Description of the event counter's register vectors.

Name	Size	Purpose
State	[1:0]	A unique number will be assigned to each state to identify it.
clock_cycles	[31:0]	Counts the number of the clock cycles elapsed since entering the COUNTING state.
events_running_total	[31:0]	Counts the events triggered by 'event_trigger' while in the COUNTING state.
events_counted	[31:0]	The value of the 'events_running_total' counter when it exited the COUNTING state, i.e. the total number of events counted in the last event counting cycle.
data_ready	[0:0]	Signals the completion of one counting cycle and the availability of new counting data.

The event counting task can now be broken down into five distinct states:

1. The STARTUP state initializes all our registers and counters.
2. The COUNTING state keeps track of two different processes. One counter increments at f_{clk} and keeps track of the number of system clock cycles elapsed, n, while being in the COUNTING state. This count value determines whether the FSM will remain in the COUNTING state (while $n < \tau_{ref}/f_{clk}$) or if it will move on to the next state (when $n = \tau_{ref}/f_{clk}$). Yet another counter keeps track of the events detected.
3. The STORE_COUNTS state stores the (running total of) events observed in a (vector) register. It corresponds to the value displayed at the output port of the module labeled 'events_counted' in the schematic in figure 5.5.
4. The RESET_COUNTERS state resets the two counters that keep track of the number of elapsed clock cycles and the running total of events. It also sets another output (not shown in the diagram above), named 'data_ready', high to signal the completion of one complete event counting cycle and the availability of new data at the module port.
5. For robustness, a RESET state exists to define the next state in case a reset event should ever occur.

Before continuing to the Verilog implementation, it will be helpful to define the various (vector) registers which will be directly affected by the FSM, see table 5.1.

5.3.3 Verilog case statements and parameters

(If it is any consolation, this is the last Verilog topic we will cover in this book!)

The Verilog implementation of the FSM relies on a case-structure such as the one shown in the example below:

```
wire [1:0] state;

//some code here to assign a value to state

case (state)
    2'b00  : begin
                // statement 0 here;
             end
    2'b01  : begin
                // statement 1 here;
             end
    2'b10  : begin
                // statement 2 here;
             end
    2'b11  : begin
                // statement 3 here;
             end
    default: begin
                // default statement  here;
             end
  endcase
```

Depending on the value of the wire vector named *state*, one and only one of the statements will be executed. (The default state is included for robustness.) This example of a case-structure acts like the 4-to-1 MUX in section 3.3.

A useful construct in Verilog is the 'parameter' keyword. It makes code more readable and easier to modify. Parameters act like constants and the compiler replaces them with their corresponding value. The example above could also have been written like:

```
wire [1:0] state;

// some code here to assign a value to state

parameter STARTUP = 2'b00;
parameter COUNTING = 2'b01;
parameter STORE_COUNTS = 2'b10;
parameter RESET_COUNTERS = 2'b11;

case (state)
    STARTUP  : begin
                // statement 0 here;
             end
    COUNTING  : begin
                // statement 1 here;
             end
    STORE_COUNTS  : begin
                // statement 2 here;
             end
    RESET_COUNTERS  : begin
                // statement 3 here;
             end
    default: begin
                // default statement  here;
             end
  endcase
```

5.3.4 FSM in Verilog and language templates

Let us implement the FSM by creating a new project with a module for the FSM. See figure 5.5 for a list of port names for the FSM module named 'SyncEventCounterV1'. (*Note*: one of its outputs, a single-bit register 'data_ready' is not shown in the schematic because we are not using it (yet).)

If we had to write the code for the entire state machine it could be daunting task. So we will employ another extremely useful Vivado tool, namely the 'Language

Template' tool, which contains Verilog code snippets. It can be opened by clicking on the 'lightbulb' icon as shown by (1) below.

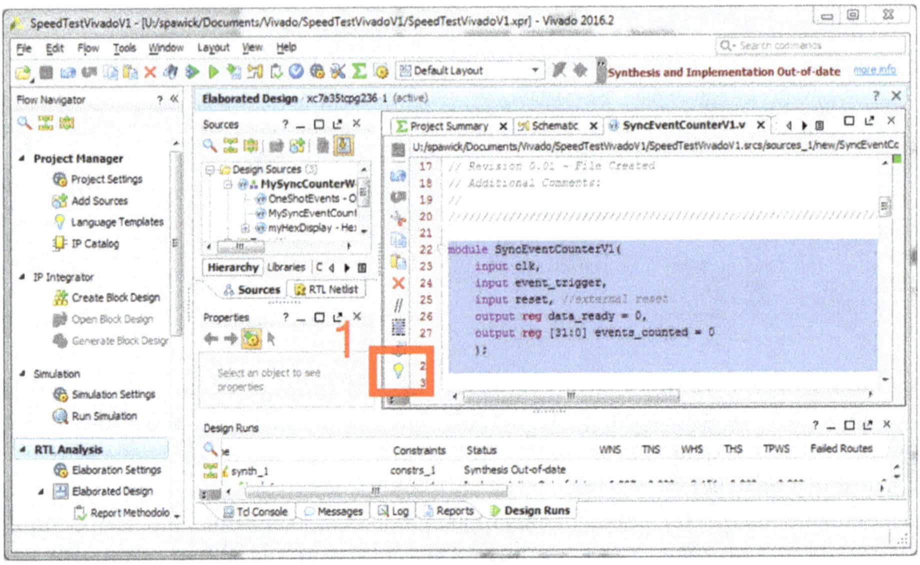

You may want to explore the 'Language Template' tool before we copy the code snippet for our FSM. Specifically, browse through the 'Verilog/Synthesis Constructs/Coding Examples' section where you will be able to find pretty much every Verilog code snippet you will ever have to write.

The FSM can be found under: 'Verilog/Synthesis Constructs/Coding Examples/State Machines/Moore/Binary (parameter)/Fast/4 States' (see the screenshot below).

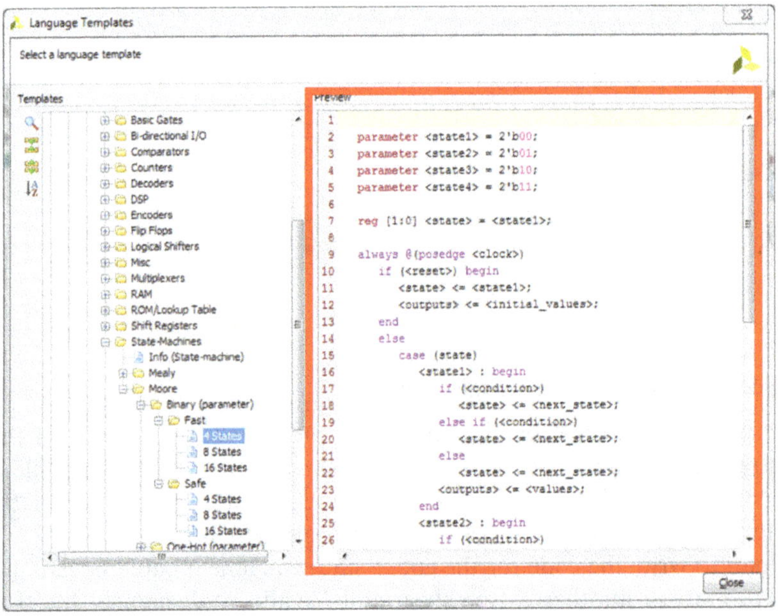

Copy all the information in the right-most window into your empty module directly before the 'endmodule' statement.

Look at the copied code snippet. While its size might be daunting, do not worry. We will delete many lines of code because they are not needed since our FSM is rather simple. Note all the signals are encased in angle brackets, such as <clock>. All of them must be replaced with names that match your actual signals, such as 'clk'. (Make sure you remove the angle brackets themselves; they are only used as placeholders.)

Before we adjust the template to our needs, notice the following three key points:

1. There are four parameterized states currently labeled <state1>, <state2>, <state3> and <state4>. You need to replace these names (throughout the entire module) with the names of the states listed above.

2. The main structure of the state machine consists of case-statements inside an always-block. Each case corresponds to one specific state. At each posedge <clock>, one and only one of the four cases (or states) listed is activated. This active state corresponds to the value which has been assigned to the register named 'state' in the previous clock cycle.

   For example, the <state1> case is listed below:

```
<state1> : begin
  if (<condition>)
      state <= <next_state>;
  else if (<condition>)
      state <= <next_state>;
  else
      state <= <next_state>;
  <outputs> <= <values>;
end
```

3. When you examine an individual case in more detail, like the one shown above, you will see that its functionality is designed for two distinct purposes:
   - to make a decision about the next state—this is accomplished by assigning a specific value to the register *state* based on the various input condition;
   - the assignment of (various) values to (various) outputs.

Also note that all assignments within the state machine are performed using sequential logic using non-blocking assignment operators and registers. This should make sense since the entire FSM is contained within a large always-block.

5.3.5 Implementing the FSM code

Your job is to modify each case from the 'Language Template' to suit the requirements for the particular state of the event counter. Below are a few clarifications/hints.

Assigning the next state. From the FSM diagram in figure 5.7 you can see that for most states there is a clear choice what the next state should be. For example, in the

RESET_COUNTERS state the assignment of the next state is independent of the input values. For such cases, you may omit the if–else statements and directly assign the next state to the 'state' register. See the example below:

```
RESET_COUNTERS : begin
    state <= COUNTING;

    clock_cycles <= 0;
    events_running_total <= 0;
    events_counted <= events_counted;
    data_ready <= 1'b1;
end
```

Assigning values to outputs. The template lists only one assignment of a value to an output: '<outputs> <= <values>;'. However, you will find that multiple such statements will be needed and, in one case, you will even need to make this a conditional assignment. See the example for the RESET_COUNTERS state shown above, where multiple such assignments are made.

(Note the seemingly redundant statement: 'events_counted <= events_counted;'. It is included to remind us how each of the five register vectors are affected by each state. The statement above is a place holder to affirm that 'events_counted' is not being altered in this particular state. Although it is redundant, it is good practice to include it.)

See if you can implement the entire FSM algorithm. For testing and trouble-shooting, make t_{ref} only ten clock cycles long (the use of a parameter for this value is highly encouraged). Use Vivado's 'Behavior Simulation' tool from the previous section to check that it counts the events correctly.

In case you have multiple modules in your Verilog project, you must set the module you are testing for the 'Behavior Simulation' as the top module in the 'Sources' window under 'Simulation Sources'. (*Note*: this module may be different from the top module under the 'Design Sources'.)

Do not bother adding the one-shot synchronizer module (yet); instead use the 'Force Clock' for both the 'clk' and the 'event-trigger' with identical values. Use the 'Force Constant' to set the 'reset' input permanently low. Finally drag the 'state' variable into the wave window and run the behavioral simulation until you have a timing diagram for one complete event counting cycle, i.e. about 20 system clock cycles. Check that 'events_counted' is 10 (or hex *a*) for these settings. Also check that 'data_ready' goes high for only one system clock cycle after executing the RESET_COUNTER state. (Save your timing diagram, you will need it for the review exercises.)

In case you get really, really stuck you may 'peek' at the complete FSM module at the end of this section but do not just copy it blindly, as this may have severe consequences in the next section where you will create your own FSM.

5.3.6 Hardware testing

Let us put all the pieces together for our event counter and then test it by sending a digital signal from the function generator into its 'asynctrigger_in' input. Start by instantiating all modules shown in figure 5.5. Make sure you assign JB[1] to the

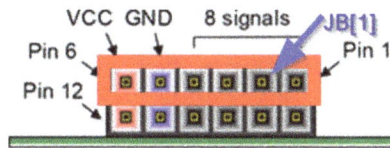

Figure 5.8. Connect the digital signal from the function generator to the Pmod B, JB[1] pin.

'asynctrigger_in' signal of the one-shot module, since that is a dedicated clock pin on the FGPA. Generate your bit-file and program the board.

Connect one of the BNC connector modules to the upper row of JB pins, i.e. the connectors on the upper right edge of the board. Make sure the blue jumper next to the BNC connector you are using is in position *B;* in the absence of such a connector feed the signal directly into the second pin from the right, in the top row, and connect your ground to the second left-most pin, as shown in figure 5.8.

Before you connect your function generator to the board, please double check the next step because it is very easy to destroy the BASYS3 board:

Make absolutely sure your function generator's output does not exceed the 3.3 V TTL level for the high output and that it does not produce negative voltages for the low output. Ideally, use the digital, sometimes labeled 'SYNC' output from your function generator.

Please double check this and verify the voltage levels on a scope before you connect the function generator to the BASYS3 board. To repeat, *under no circumstances feed an analog signal into the BASYS3 board which goes below ground or exceeds above stated high level of 3.3 V.*

Write your Verilog module so that t_{ref} corresponds to a time interval of 1 second. Set the function generator's output frequency to 1 kHz. Observe the hex display on the board while you slowly alter the function generator's frequency. Does it agree? In what units is it displaying the events? What is its range?

Change your code so that $t_{\text{ref}} = 1$ ms and feed your event counter with a faster signal. In what units is your display now? How does it agree with value set on the function generator? What is its range for $t_{\text{ref}} = 1$ ms?

Hopefully, you are impressed with your event counter's performance. Despite the fact that it is 32-bit counter, it is limited by its four-digit display. In the last section in this chapter, you will see how to interface your board with the computer and get around this annoying shortcoming.

5.3.7 Complete FSM code

If you get stuck you may 'peek' at the complete module below.

While we strongly discourage copying the code without understanding it, there is not much benefit to re-entering it and, thereby, wasting valuable time. Therefore, we have made this code available on the https://sites.google.com/a/umn.edu/mxp-fpga/home/vivado-notes/phys4051-course-related-materials site; it is named 'SyncEventCounterV1.v'.

As suggested, use this code only if you are stuck or for debugging:

```verilog
module SyncEventCounterV1(
    input clk,
    input event_trigger,
    input reset, //external reset
    output reg data_ready = 0,
    output reg [31:0] events_counted = 0
    );


    parameter STARTUP = 2'b00;
    parameter COUNTING = 2'b01;
    parameter STORE_COUNTS = 2'b10;
    parameter RESET_COUNTERS = 2'b11;

    parameter MAX_CYCLES = 100_000_000;


    reg [31:0] clock_cycles = 0;
    reg [31:0] events_running_total = 0;

    reg [1:0] state = STARTUP;


    always @(posedge clk)
        if (reset) begin
            state <= STARTUP;
        end
        else
            case (state)

                STARTUP : begin
                    state <= COUNTING;

                    clock_cycles <= 0;
                    events_running_total <= 0;
                    events_counted <= 0;
                    data_ready <= 1'b0;
                end

                COUNTING : begin
                    if (clock_cycles == (MAX_CYCLES -1) )
                        state <= STORE_COUNTS;
                    else
                        state <= COUNTING;

                    clock_cycles <= clock_cycles + 1;

                    if( event_trigger)
                        events_running_total <= events_running_total + 1;
                    else
                        events_running_total <= events_running_total;

                    events_counted <= events_counted;
                    data_ready <= 1'b0;

                end

                STORE_COUNTS : begin
                    state <= RESET_COUNTERS;

                    clock_cycles <= clock_cycles;
                    events_running_total <= events_running_total;
                    events_counted <= events_running_total;
                    data_ready <= 1'b0;
                end


                RESET_COUNTERS : begin
                    state <= COUNTING;

                    clock_cycles <= 0;
                    events_running_total <= 0;
                    events_counted <= events_counted;
                    data_ready <= 1'b1;
                end
            endcase
endmodule
```

5.3.8 Review exercises

1. Create the timing diagram you obtained from running Vivado's 'Behavior Simulation' for t_{ref} being ten system clock cycles long. Use the 'Force Clock' for both the 'clk' and the 'event-trigger' with identical values and the 'Force Constant' to set the 'reset' input permanently low. Drag the 'state' variable into the wave window and run the behavioral simulation until you have a timing diagram for one complete event counting cycle, i.e. about 20 system clock cycles.

2. Determine the range of your event counter if $t_{ref} = 1$ s and 1 ms. What units is the display in for these two cases? How do the displayed values agree with the function generator?

5.4 Period counter

5.4.1 Introduction to period counters

Frequency counters are very popular among scientists because they can produce results whose uncertainty can be reduced (theoretically) to an arbitrarily small value by simply extending the time over which the measurements are acquired.

Let us illustrate this with the following example. Consider that you are using an *ideal* event counter which works just like the one you built in the previous section. It is driven by a stable (and accurate) 100 MHz reference clock and it can display all the digits of the events recorded by its internal 32-bit counter. The input signal we are going to measure is a 'clean', periodic 1 MHz square wave.

For our first measurement, let us use a reference time interval, t_{ref}, of 1 ms. For the input signal specified earlier, i.e. 1 MHz, we would then record 1000 events in this time interval. We can easily observe any change, df, in our input signal when the display changes. For example, a change from 1000 events/ms to 1001 or 999 events/ms would signify a change in our input frequency of (at most) 1000 Hz. In other words, for our ideal event counter, in just 1 ms we can observe a percentage change, df/f, of $1000/10^6$, or 1 part in 1000.

If we repeat the experiment and increase t_{ref} to 1 s, we would then read 10^6 events. A change to 1 000 001 or 999 999 events per second corresponds in this case to a drift of 1 Hz in the 1 MHz input signal. We can see that for the 1 s t_{ref} interval the accuracy for df/f increased to 1 part in 10^6.

If we extend this thought experiment further, we could use an arbitrarily long t_{ref} yielding a df/f that approaches 0. (While this argument is correct, there is a catch: it assumes that reference time source is stable and remains completely accurate over the longer time interval.) This explains the popularity of this event counting technique because its only expense (for increasing the result's accuracy) is to wait a bit longer.

As amazing as that result is, this technique only works well for an input signal in the 'right' frequency range. So what is the 'right' or 'wrong' frequency? Let us use the same event counter as before but this time we choose an input frequency, namely 1 Hz.

With $t_{ref} = 1$ ms, we would not record any events. Therefore, the input signal would have to change by 1000 Hz to record a single event. In this case, $df/f = 1000/1$, which is a pretty lousy result.

If we change t_{ref} to 1 s, our percentage accuracy improves to 1/1. To achieve an accuracy like the 1 part in 1000 we previously obtained for our 1 MHz input signal, we would have to use a t_{ref} of 1000 s. For an even greater accuracy, such as 1 part in 10^6, we would have to count our 1 Hz events continuously for almost 12 days! This is an amazing result in itself given that our reference clock runs on a 10 ns cycle. Clearly, we are using the wrong approach to measure the 1 Hz signal.

Our technique is not working (very well) because we are measuring the event rate of a slow signal with a much faster reference clock. What we should be doing instead, is to use our fast clock to measure the time it takes for the (slow) events to repeat themselves. In other words, we should measure their period instead of their frequency and we need to reconfigure our frequency counter into a period counter.

This is how a period counter works: the reference clock counter starts when the active edge of an event is detected. It continues to count the elapsed system clock cycles until the arrival of the next active event edge. At this point, the number of reference clock cycles is stored. This value is proportional to the time interval between the two events observed, i.e. their period. The reference clock counter is reset and the (counting) cycle starts over again.

Let us return one more time to our previous thought experiment with our 1 Hz input signal. However, this time let us use an ideal period counter and calculate dt/τ where dt represents the uncertainty in the period and τ the period. (Note that t_{ref} must now be at least as long as one input signal period, since we must wait for it to complete at least one cycle.) With our system and $t_{ref} = 1$ s, we would record 10^8 system clock cycles and dt/τ would be 10^{-8}. In other words, by using the period counter, we can again achieve in just one second an accuracy which is comparable (or better) to the results obtained previously with the frequency counter with the 1 MHz input signal.

To summarize the thought experiment: for fast signals, measure their frequency; for slow ones, measure their period.

5.4.2 Exercise: implementing the period counter

It is now your job to implement such a period counter in Verilog. Let us follow the example in the previous section and start with a state machine diagram.

If you were to draw a state diagram for the period counter, you would find it would look identical to the one for the event counter shown in figure 5.7. The difference between the two designs is in the implementations of their states. Specifically, (a) how the COUNTING state is implemented, (b) which value to assign to 'counts_out' and (c) the value the reference clock counter needs to be reset to in the RESET_COUNTERS state.

Part 1. Create a copy of the frequency counter module. (Be sure to save the original; you will need it later.) Modify this module and turn it into a period counter which measures the number of system clock cycles elapsed in *one* event cycle. Once you are confident about your module, instantiate it in the top module. To test your code, use Vivado's 'Behavior Simulation' and use a 'Force Clock' signal for the 'event_trigger' input and set its frequency to 1/10 of the system clock signal, 'clk'.

Important: these 'Forced Clock' signals will only produce the correct result if they are applied to the top module inputs. Why? If you apply them instead to the period counter module directly you will also have to change the duty cycle of the 'event_trigger' signal so that it is synced to the system clock and will remain high for only one system clock cycle.

Does your simulation show that your module counted ten system clock cycles between events? (*Hint*: if you are off by two cycles think about part (c) above.)

Test your entire design with a function generator's SYNC output using a 10 kHz signal. How many system clock cycles does this period correspond to? Does your display show the correct value of system clock cycles for the 10 kHz input signal?

Part 2. The previous test verifies that your period counter works correctly. However, your BASYS3 board displays the result in system cycle or units of 10 ns, which is rather unusual. Let us fix that so that it will display the period in millisecond units.

You have two different ways to achieve this. The first method involves creating a new module which will change the system clock to a clock signal with a 1 ms period, and then feeding that signal into your existing period counter module. The second method involves extending the number of events over which your period counter counts the (original) system clock signals. Either method is fine, it is up to you.

Part 3. This part is *entirely optional* and you may do it only if you want an extra challenge: instantiate both the frequency counter and the period counter in your top module and use a MUX, controlled by a switch or button on the BASYS board, to display either the period or frequency of the input signal in milliseconds or hertz.

5.4.3 Review exercises

1. Explain the difference between a frequency counter and a period counter. Explain when it is more advantageous to use one or the other.
2. Show and explain the timing diagram of the behavior simulation of your period counter module as specified in the exercise in section 5.4.2.
3. Describe how you fixed the display issue so that it correctly displays the period in milliseconds.
4. Create the Verilog code and the elaborated schematic of your period counter.

5.5 Computer interfacing the design by embedding a microprocessor

In this last Verilog exercise, you will add a programmable microprocessor or CPU to your frequency counter design so that it communicates its results to a PC over a

USB. The exercise exposes you to another, far more powerful level of FPGA applications using an 'embedded' microprocessor in tandem with your existing design.

If you are not familiar with microprocessors, their functionality is similar to a car engine. All it really accomplishes is turning a shaft by moving a cylinder back and forth. However, the engine is what powers the car. What the engine is to a car that is what the microprocessor is to a microcontroller. The instructions a microprocessor is able to execute are extremely simple. You already worked some of them out in chapter 3 when you built part of an ALU from scratch. However, by assembling these instructions into clever algorithms and by connecting the microprocessor to peripherals, the entire system becomes a microcontroller, i.e. a computer designed to accomplish one specific task. By embedding a microprocessor into the FPGA we can now design and build custom scientific equipment from scratch.

We will use this example to lift the restriction that the BASYS board's four-digit display imposed on the range and accuracy of frequency counter's results. By being able to display the entire 32-bit frequency counter results, this application will allow us to compare the accuracy of the BASYS3 system clock to that of the commercial function generators used in the lab.

5.5.1 Data communication hardware and software

The communication hardware and software can be viewed from two distinct points, namely from the FPGA board or from the computer:

1. *Computer.* We will use a simple (dumb) terminal program (provided by Xilinx and called SDK, see below) to communicate with the FPGA. It will take care of the serial communications protocol and it will display the information received from the FPGA, i.e. the number of events recorded by our frequency counter's 32-bit counter. The data connection between the computer and the BASYS3 board will be made over the already existing USB cable which will do double duty, programming the FPGA and receiving data from it.

2. *FPGA.* To communicate with the FPGA we will implement a universal asynchronous receive and transmit (UART) circuit. Instead of building it from scratch, we will use an IP-core for a MicroBlaze MCS microprocessor which includes a UART. (We will explain IP-cores in the next section.) The serial data from the UART will be sent over the existing USB port on the BASYS3 board to the computer.

The complete schematic diagram of the project is shown in figure 5.9. It is identical to the one for the frequency counter in section 5.3, except for the addition of the microprocessor and the RsTx serial output port.

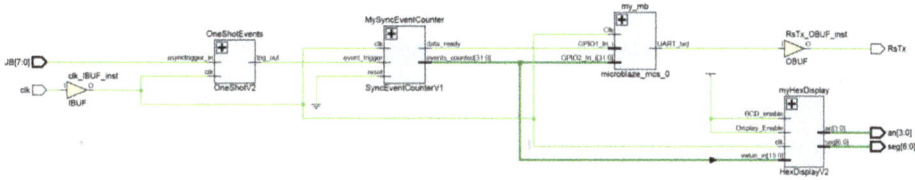

Figure 5.9. Compared to figure 5.5, the original event counter, only the MicroBlaze MCS microprocessor has been added. RsTx represent the USB UART transmission output port.

To communicate with the UART, i.e., the microprocessor, requires two different types of code and software compilers:

1. You will provide the Verilog code that implements the event counter and the microprocessor (with its internal UART).
2. You will also need code to program the microprocessor so that it knows how and when to communicate. This code will be provided to you.

Most of the Verilog code for the first part already exists from section 5.3 and it only needs some minor modifications, such as instantiating the microprocessor, connecting the event counter's output to the microprocessor's general purpose inputs (GPI) and mapping the pins for the USB UART connector in the constraint file.

However, to create and compile the code for the microprocessor we must rely on entirely new software, namely Xilinx Software Development Kit, or Xilinx SDK. It has been installed with the Xilinx Vivado as part of the Embedded Design Kit (EDK). (If you have difficulty locating it on your computer check the installation instructions in appendix A, specifically the steps enabling the additional Software Development Kit (SDK)).

Like all microprocessors, the MicroBlaze MCS is only able to process low-level machine language instructions. Working with these tends to be tedious. Luckily for us, its instructions can be written in C which is then compiled into an executable and linkable format (ELF) bit-file by the SDK. You will be provided with the C code. However, you must compile the ELF file from the code.

Finally, you will use Vivado to merge the output from Verilog and the SDK ELF file into a single bit-file that then can be uploaded to the FPGA board.

If this sounds a bit overwhelming, do not worry. We will walk you through the entire process and explain along the way what is going on. We start with the embedded design and instantiating the microprocessor. Then we will create the ELF file and finally we will generate the bit-file. Unfortunately, there are many steps involved in the process, so please bear with us.

5.5.2 Embedded designs and IP-cores

Embedded designs refer to merging the 'FPGA fabric' digital designs, such as your frequency counter, with a programmable microprocessor. The microprocessor may exist as a hardware component already inside of the FPGA or, as in our case, is loaded as an HDL implementation. What this implies is that 'someone' (or rather a team of engineers) has sat down and designed a Verilog module that contains all the functionality of a typical microprocessor. (Are not you lucky we did not assign this task to you?) Your job will be to specify the features of the microprocessor you would like to implement and then to instantiate the design, a task you should be familiar with.

The microprocessor design exists in what is called an IP-core, short for 'intellectual property' block. These IP-cores are to HDL what subroutines or libraries are to other programming languages. In other words, while they rely on the basic (digital) structure of HDL, they allow the user to implement sophisticated and complex tasks such as signal processing, filtering, data communication, mathematical operations and microprocessors on an FPGA by instantiating the IP-core. The IP-cores allow you to transform your simple FPGA into a sophisticated instrument.

Xilinx does provide some free IP-cores with Vivado. If you cannot find what you need, Xilinx and other vendors will create one for your application, although probably not for free. In other words, the business of creating and selling IP-cores is another important aspect of the entire FPGA concept.

Step 1. Creating and configuring the microprocessor

To explore the types of free IP-cores provided with Vivado, start with a *working* copy of your frequency counter project from section 5.3. (Please do *not* use your period counter from section 5.4 because the fast refresh rate may overload the data transmission process to the PC.) Also check that your frequency counter's t_{ref} is again set to 1 s.

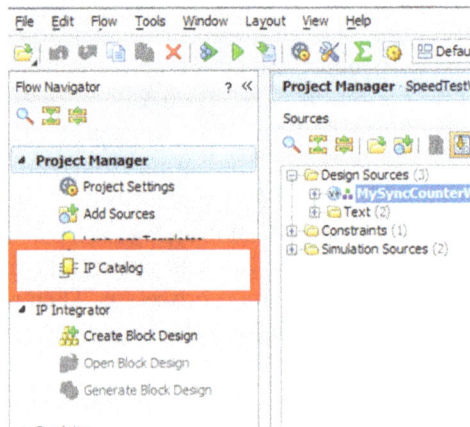

Begin by clicking in the Project Manager window on the 'IP Catalog' as shown above.

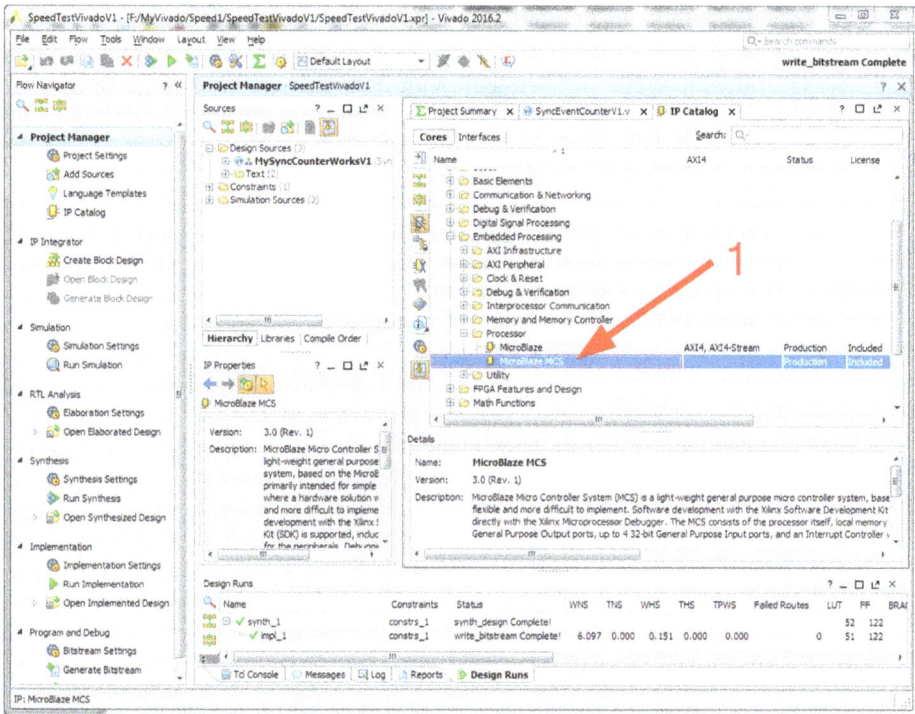

The 'IP Catalog' window (shown above) opens and displays the IP 'Cores' available for your FPGA, i.e. the Artix-7. You may want to glance at all the options available to you. However, for now you want the IP-core named 'MicroBlaze MCS' under the 'Embedded Processing/Processor' cores. (Do not confuse it with the similarly named, much more sophisticated MicroBlaze processor directly above!) Double-click on the MicroBlaze MCS (1). A 'Customizing IP' window briefly appears and then the configuration window shown below opens.

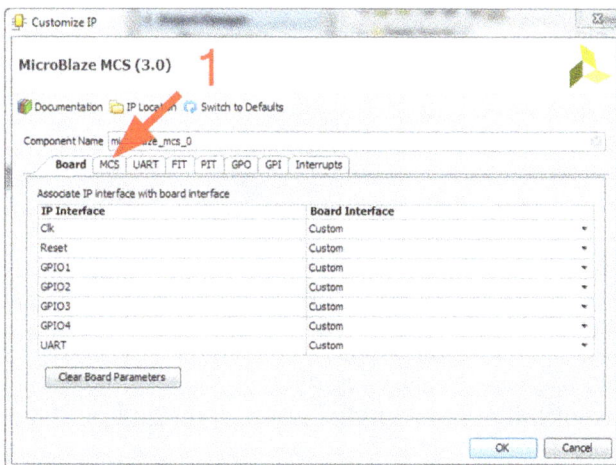

This window lists the associated documentation for the IP-core and all its customization options. Start by configuring the microprocessor's size. Click on the 'MCS' tab (1 above) and the window below opens.

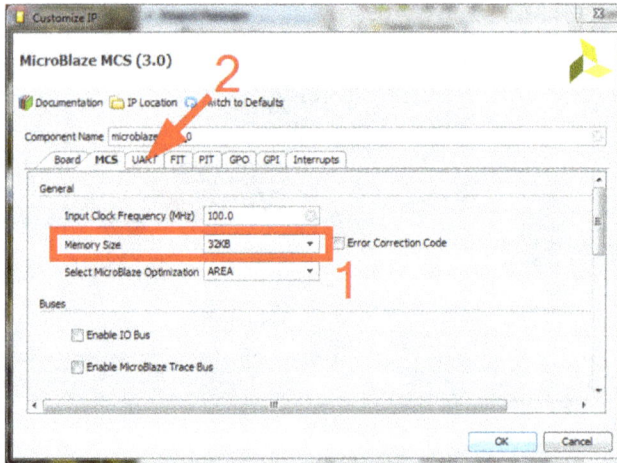

Increase the microprocessor's memory to 32 KB so that it has enough memory to store its embedded program code (1). Next select the UART tab (2) to configure the communication parameters to the values shown by boxes 1 and 2 below. These are typical communication parameters. Since we are not sending any data from the computer to the board we are not going to enable the receiver.

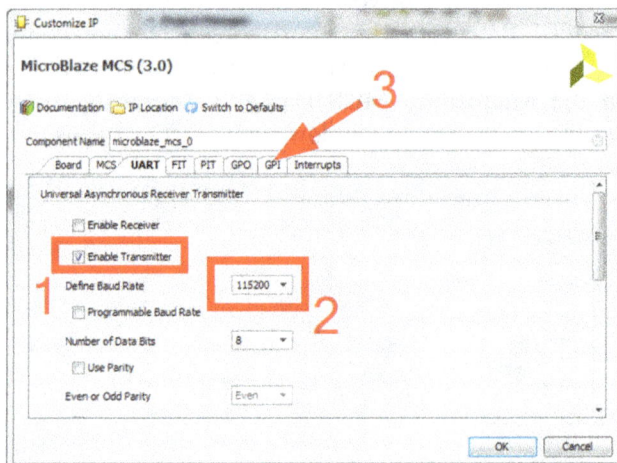

In the last step (3), we configure the microprocessor's 'General Purpose Input' registers which will be directly connected to the outputs of our frequency counter.

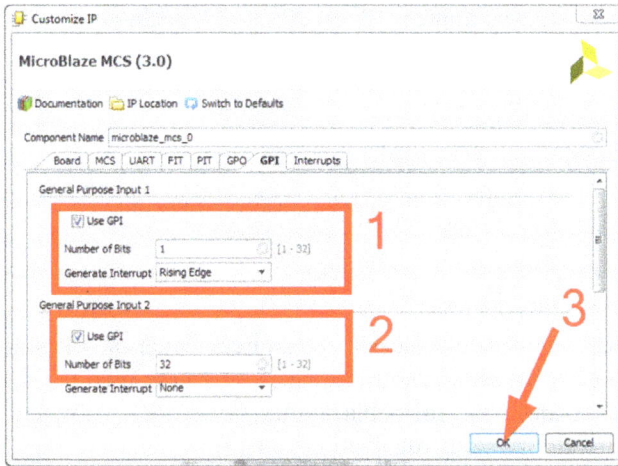

We will use 'General Purpose Input 1' (GPI1) for our 'data_ready' signal (1 above). Make sure you select the 'Rising Edge' under 'Generate Interrupt'. This way, each time it goes high, the microprocessor will generate an interrupt—which is a fancy way of saying that it will run a small program—signaling to the UART that new data are available for transmission. The actual data transmitted are the counts from the 'events_counted' output; these will be sent to the 'General Purpose Input 2' (GPI2), (2). Adjust the number of bits for each of those signals as shown above. Click 'OK' when you have made all these changes.

Vivado will now churn for a while and then starts creating the entire HDL code for the microprocessor per your specifications.

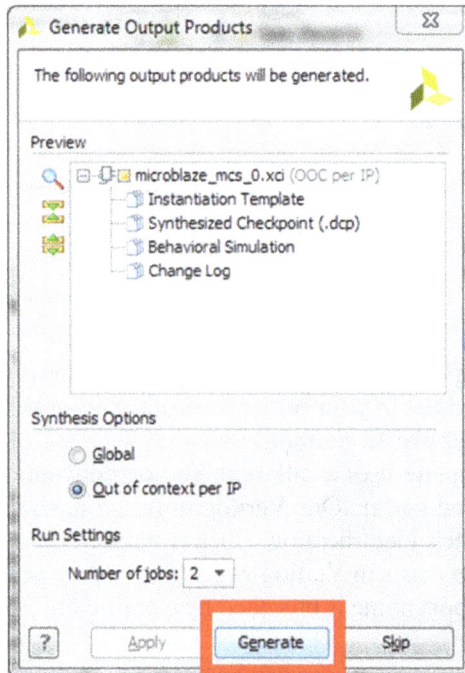

A few more nag screens appear.

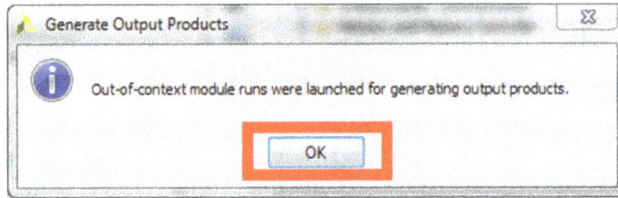

Once the process completes, click in the sources panel on 'IP Sources' as shown below (1). (If you cannot see the sources panel, go to the menu and select 'Window \Sources'.) Double-click on 'microblaze_mcs_0.veo' (2) to reveal the complete instantiation template for your microprocessor (3).

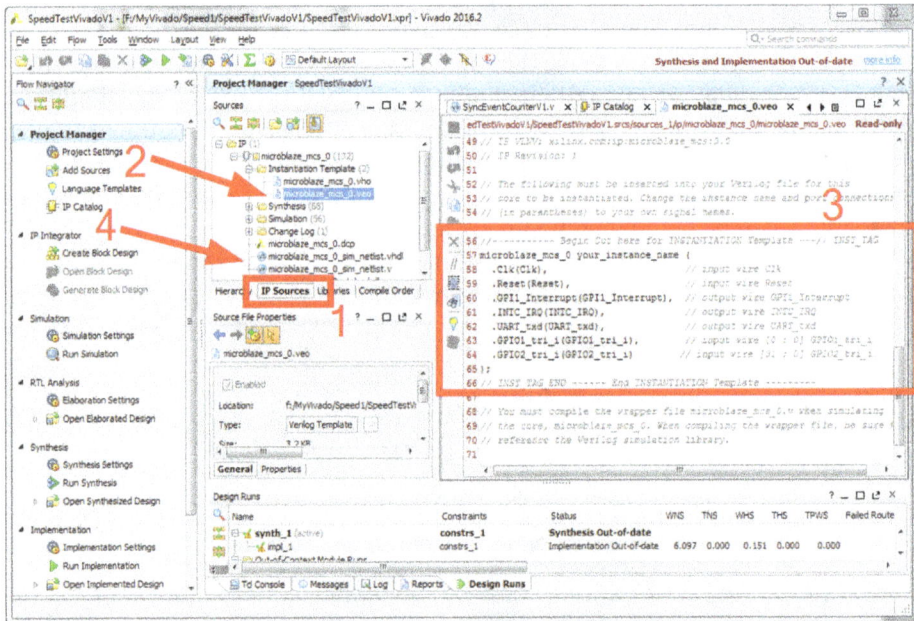

Click on the Verilog code in box 3 above and copy it into your top module. (To return to the list of modules in your project, select the 'Hierarchy' tab next to the 'IP Sources' tab, 1 in the above screenshot.)

Notice that the template uses a different Verilog convention for listing its ports than what we have used so far. Our Verilog notation has relied on preserving the order of the ports for their identification which is standard in mathematical notation. An alternative notation exists in Verilog which does not depend on the ports' order. Instead, each original port name is preceded by a period and then the new names are included within parentheses. For example, a simple module that was declared as 'MyMod(input shrek, input fiona, output donkey);' could be instantiated in this notation as: 'MyMod MyInstMod(.fiona(you), .shrek(me), donkey(him));'. This notation is preferred for large numbers of ports and also when you do not care about the state of some ports; in this case you can leave the argument blank as in 'MyMod MyInstMod(.fiona(you), .shrek(me), donkey());'.

If you have used the suggested port names in the previous exercises than your instantiation template should look like the one below. If not, adjust accordingly and verify that the wires, whose names have been highlighted in bold, all correspond to existing wires in your top module. (At this point you may change the name of your instantiated module; in the example below it has been changed to 'my_mb', although it really does not matter.)

```
//----------- Begin Cut here for INSTANTIATION Template ---// INST_TAG
microblaze_mcs_0 my_mb (
  .Clk(clk),                      // input wire Clk
  .Reset(reset),                  // input wire Reset
  .GPI1_Interrupt(),  // output wire GPI1_Interrupt
  .INTC_IRQ(),             // output wire INTC_IRQ
  .UART_txd(RsTx),            // output wire UART_txd
  .GPIO1_tri_i(data_ready),       // input wire [0 : 0] GPIO1_tri_i
  .GPIO2_tri_i(events_counted)        // input wire [31 : 0] GPIO2_tri_i
);
// INST_TAG_END ------ End INSTANTIATION Template ---------
```

Note: During the instantiation of the microprocessor you have created a new wire, named 'RsTx'. It facilitates the data communication between the USB port and the microprocessor. However, it can only do so if it has been declared in the top module's port list as an output port, as shown below:

```
module MySyncCounterwUART(
    input clk,
    input [7:0] JB,

    output RsTx, //UART Port, add to port list of your top module

    output [6:0] seg,
    output [3:0] an
);
```

Add the 'RsTx' port to your top module and uncomment lines affiliated with the 'RsTx' port in your constraint file. (Double-check these two items as they are easily overlooked!)

Check your code and confirm that that the 'events_counted' vector is 32 bits wide (everywhere) and that its t_{ref} is 1 s, i.e. that it will display its results in hertz. (*Hint*: check parameter MAX_CYCLES.)

Run 'Synthesis' only—do not proceed further, i.e. do not (yet) run the implementation.

Before we move to the Software Development Kit (SDK) we need to export the information about our hardware specifications for our microprocessor. From the main menu select 'File/Export/Export Hardware'.

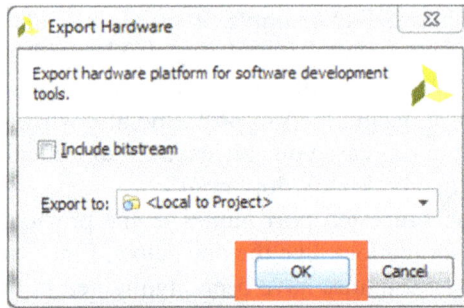

We are now done with Verilog (for a while) and move on to the SDK. Select from the main menu 'File/Launch SDK'.

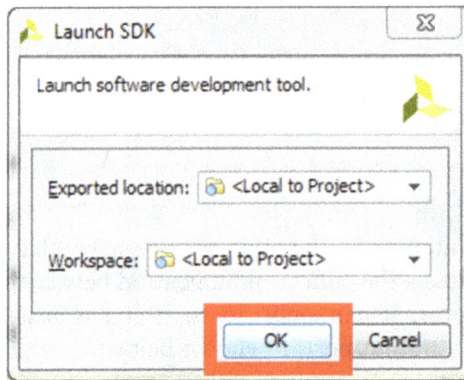

Step 2. Xilinx Software Development Kit to create the ELF file

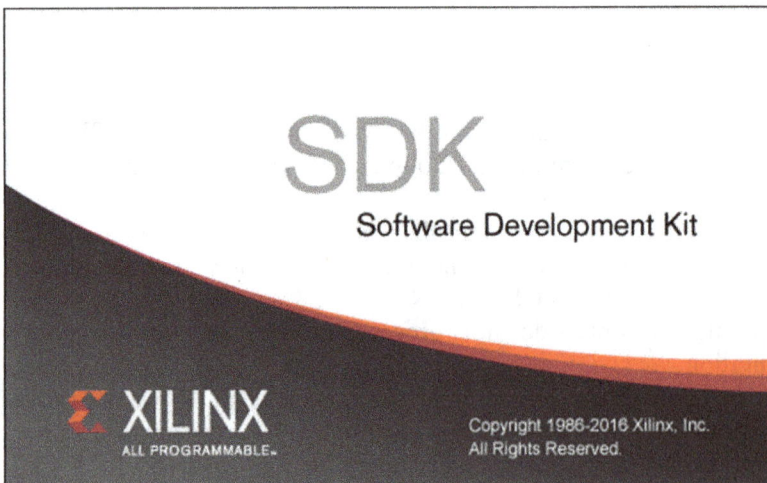

You will use the SDK environment to compile the code that controls our micro-processor. You will create an 'Application Project' and then replace its default generic 'helloworld'c.' file with an identically named file that contains the C-code for reading GPIs and transmitting the data to the UART. SDK will generate the executable and linkable format (ELF) file which you will then export into Vivado for the final bit-file generation.

Once SDK has been successfully launched, its IDE opens as shown below, or select 'Create Application Project'.

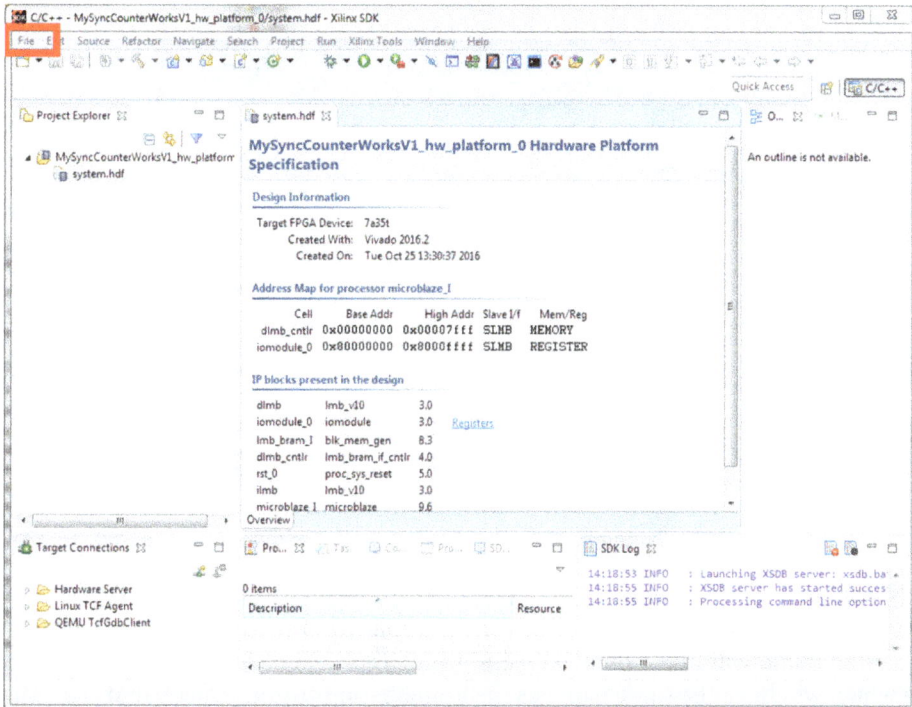

Start a new application project by selecting from the main menu 'File/New/ Application Project'. The window below opens.

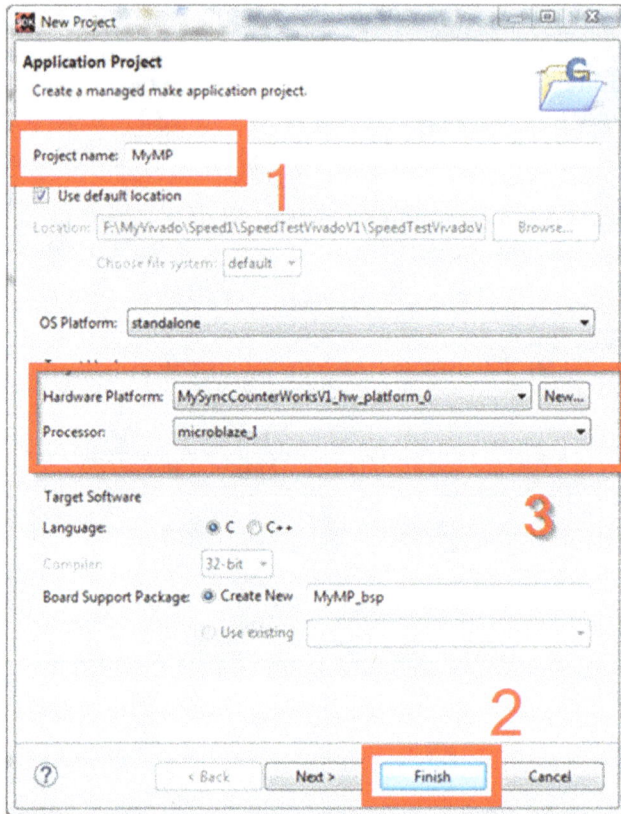

Name your application project 'MyMP' (1 above) and press 'Finish' (2). (*Note*: you could choose any name you like for the application project; however, using the suggested name will make the next steps a bit easier when we have to find the ELF file which will have the same name as the application project.) Check that the values in box 3 agree with your project, i.e. that it corresponds to your Verilog top module's name appended by '_hw_platform_' and that it lists the MicroBlaze processor; adjust if necessary.

Now comes the part where we replace the default 'helloworld.c' file with our file which has (confusingly) the same name. (The original file was created to send a simple 'Hello World' message to the UART to be used for debugging purposes.)

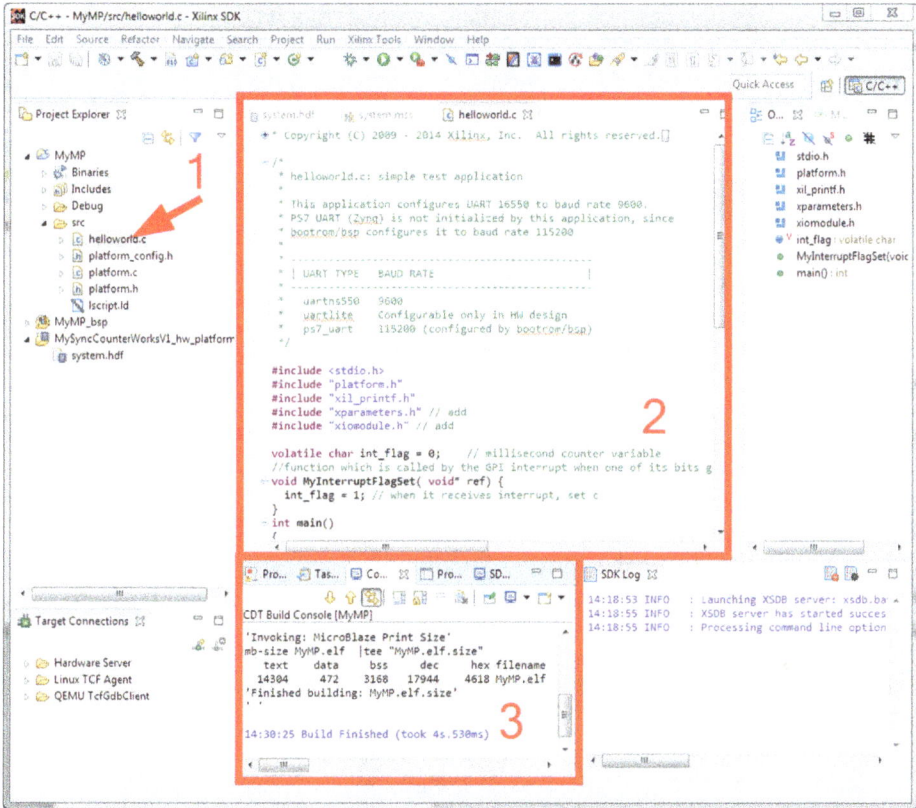

Back in SDK's 'Project Explorer', right-click on the 'helloworld.c' file and delete it (1). It can be found in the 'src' folder under the 'MyMP' application project.

Next move the new 'helloworld.c' file from the https://sites.google.com/a/umn.edu/mxp-fpga/home/vivado-notes/phys4051-course-related-materials folder into the same location, i.e. drag and drop or copy/paste it directly on the 'src' folder under the 'MyMP' application project (1). When the message below appears, click 'OK'.

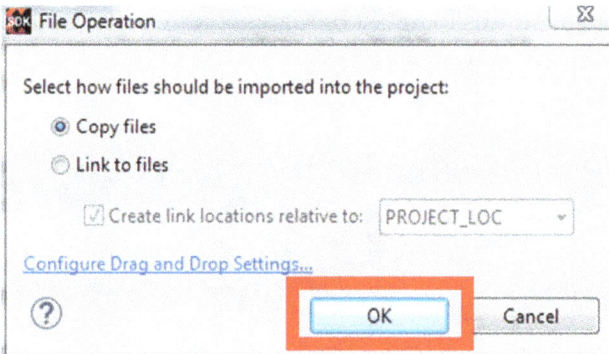

Double-click on the newly imported helloworld.c' file; its contents should appear like the one shown in panel 2 in the screenshot previous to the above.

You may want to briefly glance at the code. It instructs the processor to set a 'flag' each time it receives an interrupt. As you specified earlier when you instantiated the microprocessor, this interrupt is caused by the 'data_ready' signal. The first part of the main program consists of configuring and enabling the interrupts and the GPI inputs. It also contains an infinite while-loop which 'prints' the elapsed seconds and the number of counts received by the GPI2 input to the UART. This loop advances each time an interrupt has been received.

The ELF file is created automatically each time you save the project. So save all your files and check that the SDK environment's 'Console' window displays the 'Finished Building: MyMP.elf.size' message, as shown by 3 in the screenshot above.

Leave SDK open and return to Vivado to associate the ELF files with your project.

Step 3. Associating the ELF files with your Vivado project

Back in Vivado, click on the 'Project Manager' in the 'Flow Navigator' window (box 1 in the screenshot below) and then select from the top menu 'Tools/Associate ELF files...' (2).

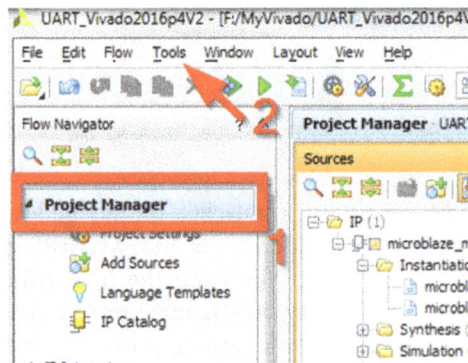

The window below opens followed by yet another window shown below.

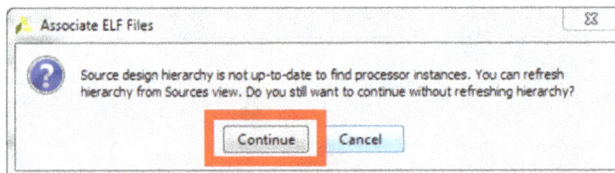

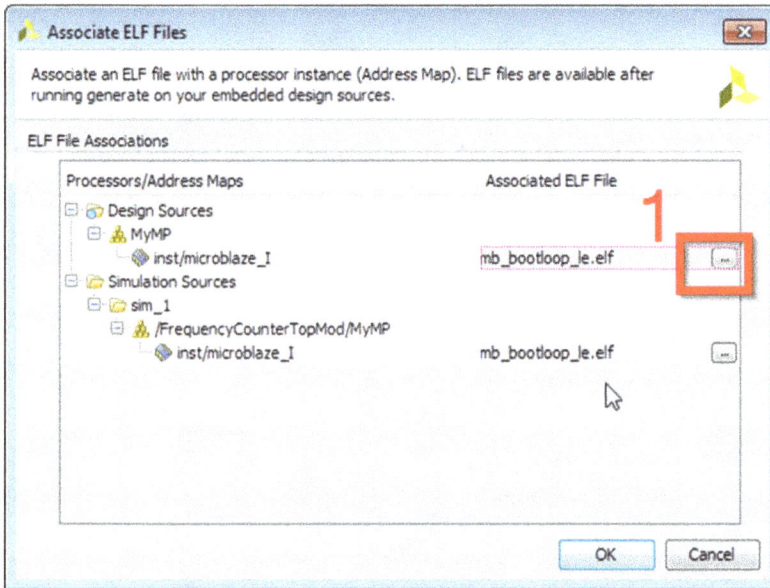

This window shows that your current Vivado project uses the default 'mb_boot-loop_le.elf' file. We need to replace it with our newly created 'MyMP.elf' file.
  Select the browse button (1), which opens the window below.

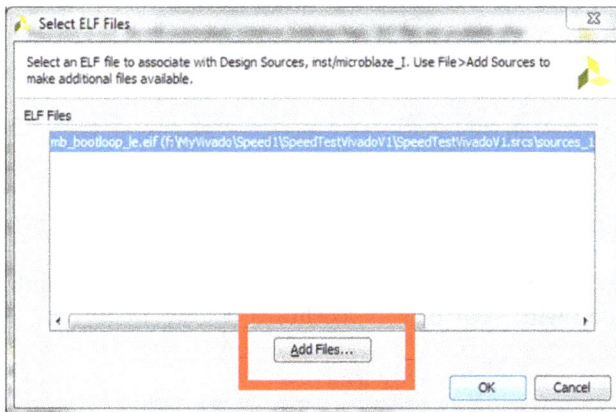

Press the 'Add Files' button and locate the newly created 'MyMP.elf' file. It can be found in your project directory in the subfolder 'PROJECTNAME.sdk/MyMP/ Debug' directory. Double-click on the file name (1 below).

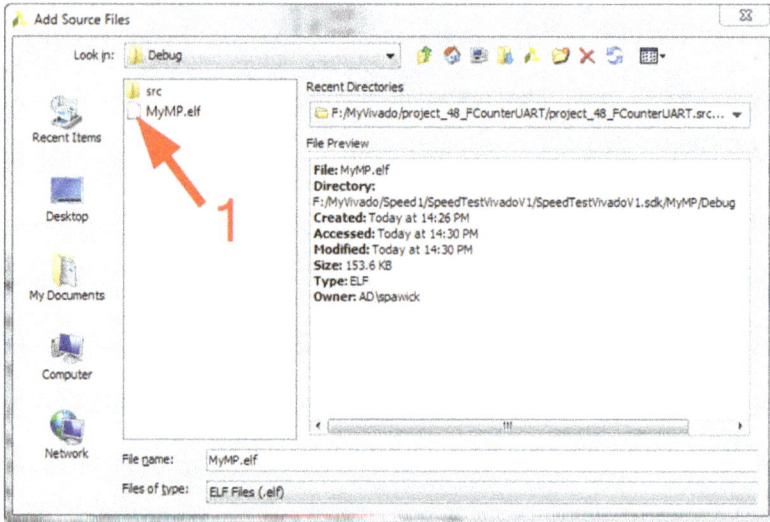

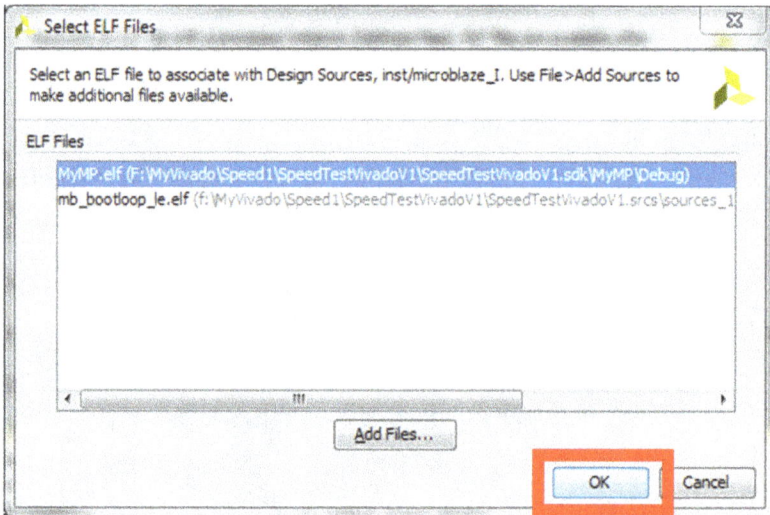

Hit 'OK' after the window shown below opens. You should finally see the new ELF file associated with your project as shown below.

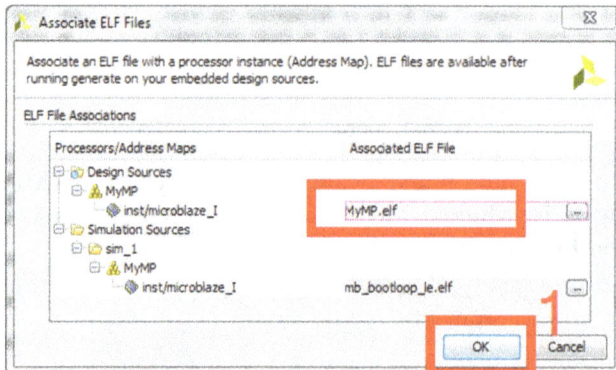

Step 4. Generate the bit-file

We are finally done. Run 'Implementation' and 'Generate the Bitstream' and program the board. Connect the function generator's SYNC output to the JB port on your BASYS3 board. Again, set the blue jumper next to the BNC connector to B since your input signal is sent to pin JB[1].

Step 5. Launch the communications program

Let us go back to Xilinx SDK and select its 'SDK Terminal' program to display the data that are being sent via the USB cable to the computer. (If you prefer you may use any other communication program you like, such as for example SecureCRT.) Click on the 'SDK Terminal tab' (box 1 below).

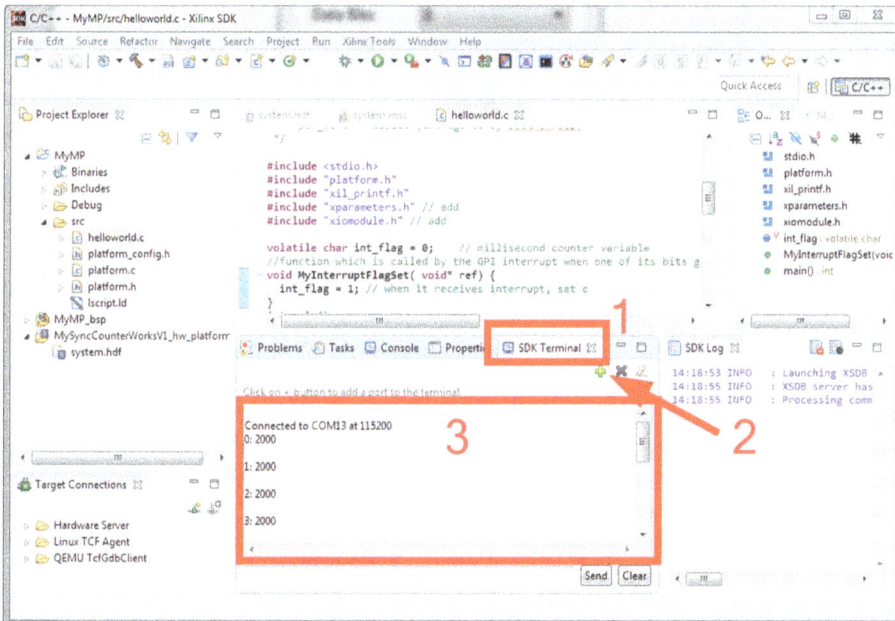

Next open a communications port by clicking on the plus sign (2 above). The window below opens and asks you to choose a communication port. The Windows operating system automatically assigns these to the USB ports so it can be sometimes difficult to find the correct port. However, starting with the highest numbered port is usually a good choice. (*Note*: this may or may not correspond to COM13 on your particular machine.)

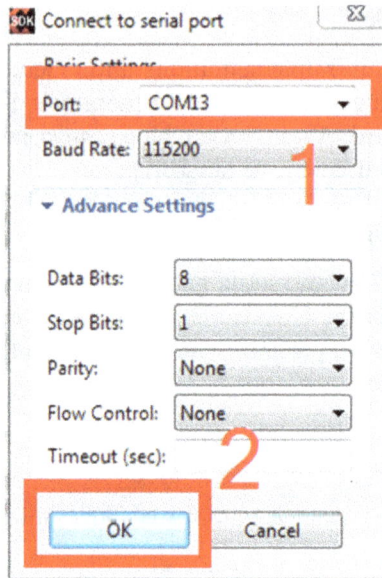

When, or if, everything works as it should, you will see the time elapsed since the board was started and the counts from your frequency counter (box 3 in the screenshot previous to the above).

5.5.3 Final exercise

The exercise is rather simple: since we are no longer constrained by the BASYS board's four-digit display, let us check out the performance of our frequency counter with very fast input signals in the megahertz range. Send the fastest signal your function generator is capable of producing to your frequency counter. How do the counts compare with the frequency the function generator displays? What is the relative error with respect to each other? Which one do you trust, the function generator or the BASYS3 board, and why?

5.5.4 Debug suggestion

If you have trouble getting this exercise to work, try the following approach: start at the beginning of this section, 5.5. Use the supplied 'SyncEventCounterV1.v', 'OneShotSyncV4.v' and the 'HexDisplayV2' files and instantiate those in your new top module, as shown in figure 5.5. Add the 'BASYS3_Master.xdc' constraint file and uncomment the appropriate lines. Program the BASYS3 board. Before you continue, confirm that the board correctly displays input frequencies applied to pin JB[1] of up to 9999 Hz.

Next follow the instructions in section 5.5, starting with step 1. Stop when you get past the third screenshot of step 2, at the paragraph starting with 'Now comes the part.' At this point, do not delete or replace the 'helloworld.c' file; instead use the one originally created by SDK. Next skip directly to the start of step 3 and continue with the remaining instructions to the end of this section. Now, each time you program the board with your bit-file, you should see a simple 'Hello World' message displayed in box 3 of the first screenshot of step 5. If you do not see this message then you have an issue with your USB connection because: (a) you may not have declared the 'RsTx' pin in your top module's port list, (ii) you may have forgotten to uncomment the 'RsTx' lines in the constraint file, or (c) you may have selected the wrong communication port or communication parameters in the second screenshot of step 5; select a different COM port number and program the board again. Only when you see the 'Hello World' message, go back and exchange SDK's 'helloworld' file with the one supplied, as outlined in the instructions in step 2, after the third screenshot. Recompile everything and program your board again. Hopefully, it will now work.

5.5.5 Review exercises

1. Explain what an embedded design is and why you would want to use one. Also briefly explain what IP-cores are.
2. Show your elaborated schematic and your Verilog code for this exercise.
3. By what value did the function generator and the frequency counter's result differ when you set the function generator to its highest frequency? What was this frequency? Which one do you trust, the function generator's or the BASYS3 board's frequency, and why? How stable were the results over a minute or so?

Appendix A

Hardware and software requirements

A.1 Software requirements

The exercises in this book use Xilinx's Vivado HL WebPACK to compile the Verilog code into a bit-file that is then used to program the FPGA on the Digilent BASYS3 board. This software can be downloaded for free directly from Xilinx at Xilinx.com. Both Windows and Linux versions exist; however, there is no Mac OS version.

For this book we have used exclusively Vivado HL WebPACK Edition 2016.2. We have tested some of the exercises with the newer version 2016.4 and have not found any problems.

A.1.1 Xilinx's Vivado HL WebPACK download and installation instructions

The free Vivado HL WebPACK software can be downloaded from Xilinx's website at www.xilinx.com/support/download.html.

Prior to downloading, you must create a free user account. Locate the Vivado HL WebPACK Edition 2016.4 download link and start the download. (See the screenshot below for the available versions.)

Warning: Vivado requires a large amount of disk space: the download alone requires 5.4 GB, the completed installation is about 15 GB.

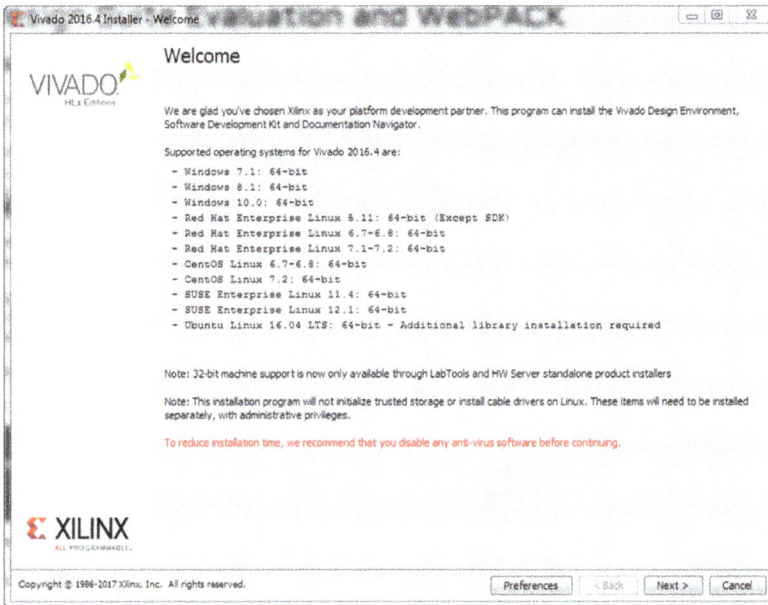

After entering your user name and password, and agreeing to all the software terms, select the free Vivado HL WebPACK software.

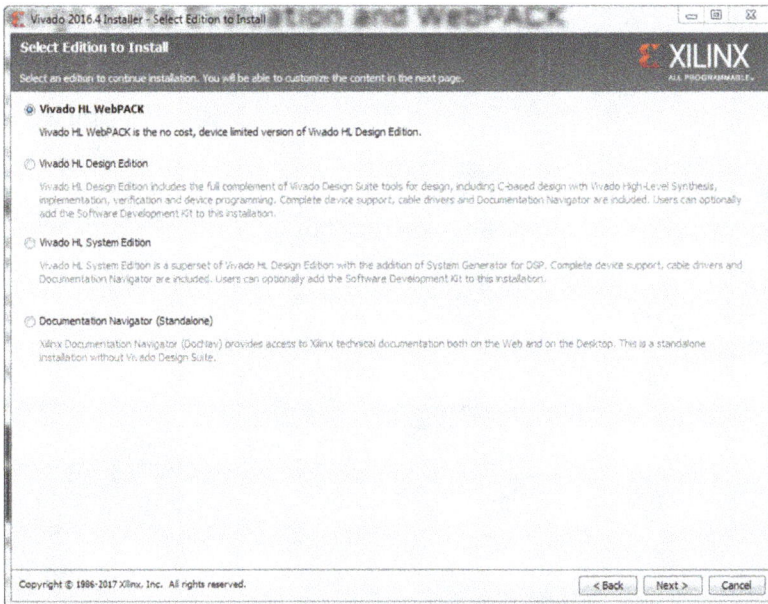

Enable the additional Software Development Kit (SDK) option in your customization option, see the screenshot below.

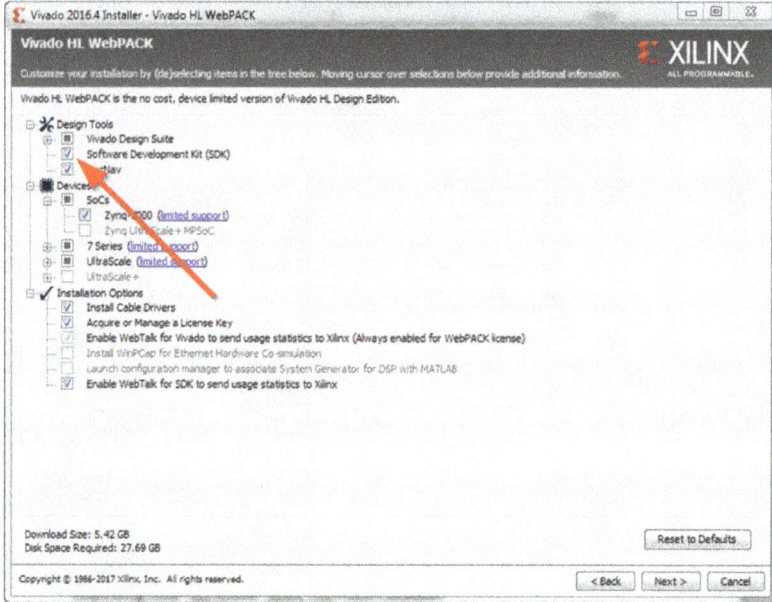

Finally, install the license by requesting the free license by email.

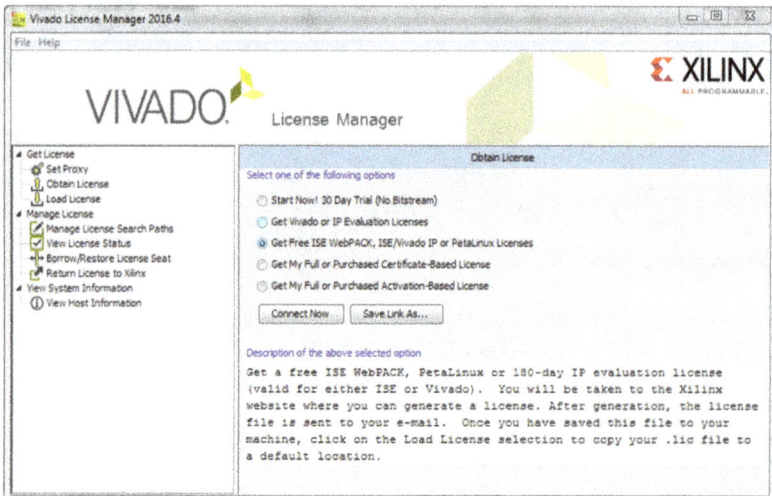

Copy the attached license into any folder and then activate it using the 'Copy License' button as shown below.

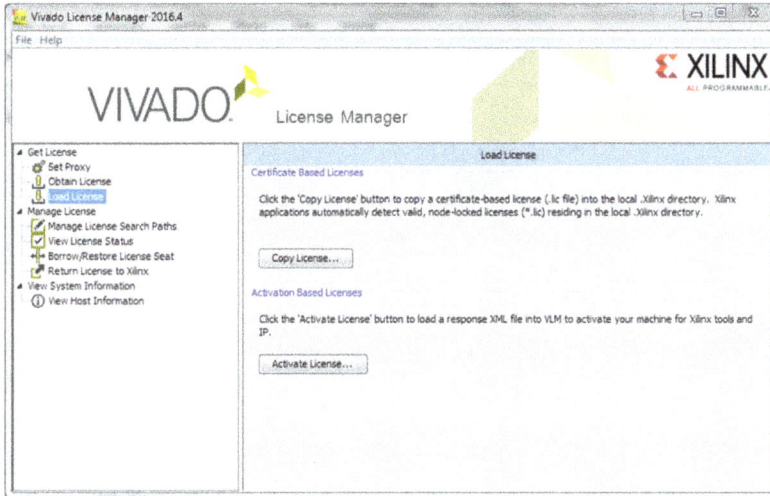

A.1.2 BASYS3 board support files (optional)

The Xilinx compiler requires detailed hardware specifications about the FPGA to be programmed. For example for the FPGA on the BASYS3 boards you must provide the following: Xilinx Artix-7 FPGA, part number xc7a35t in a CPG236 pin package with a speed grade of −1.

This information must be entered at the start of each new Vivado project which can be tedious. However, if we install the 'Board Support Package Files' we can simply select a board, such as the BASYS3, and the files will take care of supplying all the details. Installing these files simplifies the project creation process and avoids mistakes. However, the installation of these files is optional. (They are extremely useful if you decide to explore the 'Block Design' features in Vivado.)

You can find the board support files at Digilent's website:

https://reference.digilentinc.com/reference/software/vivado/board-files?redirect=1.

You may follow the installation instructions on the website listed above, or the ones provided below:

1. Unzip the files enclosed.
2. Locate the 'Vivado-boards-master/new/board_files' folder and select and copy the entire folder.
3. Locate the installation folder for Vivado, usually named 'Xilinx'. Search its subfolder until you find the subfolder named 'boards' at '\Xilinx\Vivado \2016.X\data\boards' (where X stands for your downloaded version). Now

copy/merge the 'board_files' from the zip files with the existing 'board_files'. In the 'board_files' folder, you should ultimately see the 13 subfolders as shown below, one of which will be for the BASYS3 board.

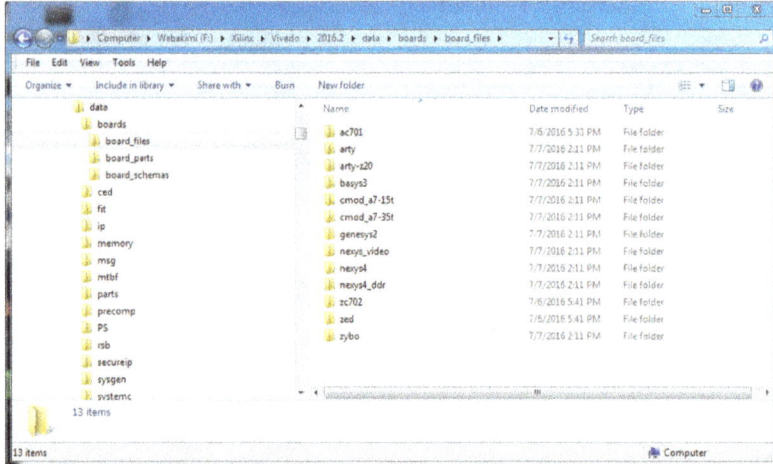

A.1.3 Master constraint file

The master constraint file maps specific pins of the FPGA with the physical devices on the BASYS3 board, such as the switches, LEDs and input and output ports. It can be downloaded from:

https://github.com/Digilent/Basys3/tree/master/Resources/XDC.

A.1.4 Other Software

Finally, you may find some of the files and modules used in this book on our Google page at:

https://sites.google.com/a/umn.edu/mxp-fpga/home/vivado-notes/phys4051-course-related-materials.

A.2 Hardware

A.2.1 FPGA

For this book we utilized a BASYS3 board from Digilent Inc., which contains an Artix-7 FPGA. It can be purchased directly from the manufacturer or other electronic suppliers. The reference manual and schematic can be downloaded from Digilent's website:

https://reference.digilentinc.com/_media/basys3:basys3_rm.pdf.

The detailed instructions are written specifically for the BASYS3 board; other boards with a suitable Xilinx FPGA may also work as long as they are supported by Vivado. (*Warning*: earlier versions of the BASYS boards, such as the BASYS and BASYS2 will not work with Vivado and will instead require an older software package from Xilinx called ISE Webpack, which differs significantly from Vivado's interface.) If you should use a different board then you must adjust the pin mapping of the FPGA to the board components accordingly.

A.2.2 Other hardware

It will be helpful to have access to a digital oscilloscope and a function generator.

www.ingramcontent.com/pod-product-compliance
Lightning Source LLC
Chambersburg PA
CBHW081535220326

41598CB00036B/6447